O-GlcNAcylation and Cancer

O-GlcNAcylation and Cancer

Editors

Ikram El Yazidi-Belkoura (Editor)
Tony Lefebvre (Co-Editor)

MDPI • Basel • Beijing • Wuhan • Barcelona • Belgrade • Manchester • Tokyo • Cluj • Tianjin

Editor
Ikram El Yazidi-Belkoura
Biology, CNRS, UMR 8576 -
UGSF - Unité de Glycobiologie
Structurale et Fonctionnelle,
F-59000
Université de Lille
Lille
France

Co-Editor
Tony Lefebvre
Biology, CNRS, UMR 8576 -
UGSF - Unité de Glycobiologie
Structurale et Fonctionnelle,
F-59000
Université de Lille
Lille
France

Editorial Office
MDPI
St. Alban-Anlage 66
4052 Basel, Switzerland

This is a reprint of articles from the Special Issue published online in the open access journal *Cancers* (ISSN 2072-6694) (available at: www.mdpi.com/journal/cancers/special_issues/O_GlcNAcylation).

For citation purposes, cite each article independently as indicated on the article page online and as indicated below:

LastName, A.A.; LastName, B.B.; LastName, C.C. Article Title. *Journal Name* **Year**, *Volume Number*, Page Range.

ISBN 978-3-0365-3083-3 (Hbk)
ISBN 978-3-0365-3082-6 (PDF)

Contents

About the Editors

Ikram El Yazidi-Belkoura is a professor in the Department of Biology at the Faculty of Sciences and Technologies of the University of Lille. Pr El Yazidi-Belkoura received a doctorate in Life and Health Sciences for her seminal contributions on the roles of molecular mechanisms of Fibroblast Growth Factor in breast cancer. From 1996 to 2010, on the basis of expertise in proteomics, Pr El Yazidi-Belkoura focused her work on growth factors' (FGFs, NGF) signaling and implication in growth/cell survival and invasion in cancer. In particular, Pr El Yazidi has participated in demonstrating the proliferating and the anti-apoptotic role of Fibroblast Growth Factors and Nerve Growth Factor in breast cancer cells under chemotherapeutic drugs. Pr El Yazidi-Belkoura's current research is focused on the role of protein O-GlcNAcylation in the chemotherapeutic response and mitogenic signaling pathways of colon cancer. The study of the impact of this post-translational modification on the regulation of Akt/mTOR signaling pathway shows that crosstalk exists between *O*-GlcNAcylation and mTOR signaling in contexts of metabolism dysregulation associated with obesity or cancer. The analysis of the role of *O*-GlcNAcylation in chemosensitivity of colorectal cancer to 5-Fluorouracil leads her to highlight that thymidylate synthase (TS) *O*-GlcNAcylation is a molecular mechanism of 5-FU sensitization and suggests combining 5-FU based chemotherapy with *O*-GlcNAc modulators. Future research on the link between patient TS gene status and *O*-GlcNAcylation dysregulation in cancer may converge towards effective personalized treatments.

Tony Lefebvre is Professor in Biochemistry in the Unit of Structural and Functional Glycobiology at the University of Lille (France). He was one of the pioneers in the *O*-GlcNAcylation area in Europe and he now leads the team "Regulation of molecular and mechanisms by *O*-GlcNAcylation". The central position of *O*-GlcNAc in metabolism has allowed his team to see a whole new field of regulation and molecular derangement related to nutrition, with cancer being placed at the forefront. His national and international connections provide access to a whole range of expertise and knowledge helping to develop his research projects. A biochemist and fundamentalist by training through research, his research subjects have gradually taken on a more applied role, which allows him to be at the interface of the academic, medical and therapeutic, and private sectors.

Preface to "*O*-GlcNAcylation and Cancer"

Cancer is one of the leading causes of morbidity and mortality worldwide, linked in part to deteriorating environmental conditions and unsafe lifestyles. Much progress has been made in recent decades in deciphering the etiology, origin, and causes of cancer, but much remains to be done to understand the molecular mechanisms underlying its initiation and progression.

In recent years, *O*-GlcNAcylation has been found to be particularly disturbed in oncogenic processes, irrespective of the stage or molecular process considered. In particular, because of its strategic position as a metabolic hub, this post-translational modification could play the role of a switch between nutritional disorders and cancer. This Special Issue is an opportunity to take stock of the current knowledge on the molecular mechanisms of cancer to use as a foundation for research in the near future, which should shed new light on our understanding of the impact of shaken *O*-GlcNAcylation homeostasis, related or not with nutritional and environmental aspects, on cancer and anticancer therapies.

Ikram El Yazidi-Belkoura, Tony Lefebvre
Editors

cancers

MDPI

Review

O-GlcNAcylation and O-GlcNAc Cycling Regulate Gene Transcription: Emerging Roles in Cancer

Matthew P. Parker [1], Kenneth R. Peterson [1,2,3,*] and Chad Slawson [1,3,*]

1 Department of Biochemistry and Molecular Biology, University of Kansas Medical Center, Kansas City, KS 66160, USA; mparker6@kumc.edu
2 Department of Anatomy and Cell Biology, University of Kansas Medical Center, Kansas City, KS 66160, USA
3 KU Cancer Center, University of Kansas Medical Center, Kansas City, KS 66160, USA
* Correspondence: kpeterson@kumc.edu (K.R.P.); cslawson@kumc.edu (C.S.)

Simple Summary: O-linked β-N-acetylglucosamine (O-GlcNAc) is a post-translational modification (PTM) linking nutrient flux through the hexosamine biosynthetic pathway (HBP) to gene transcription. Mounting experimental and clinical data implicates aberrant O-GlcNAcylation in the development and progression of cancer. Herein, we discuss how alteration of O-GlcNAc-regulated transcriptional mechanisms leads to atypical gene expression in cancer. We discuss the challenges associated with studying O-GlcNAc function and present several new approaches for studies of O-GlcNAc-regulated transcription.

Abstract: O-linked β-N-acetylglucosamine (O-GlcNAc) is a single sugar post-translational modification (PTM) of intracellular proteins linking nutrient flux through the Hexosamine Biosynthetic Pathway (HBP) to the control of cis-regulatory elements in the genome. Aberrant O-GlcNAcylation is associated with the development, progression, and alterations in gene expression in cancer. O-GlcNAc cycling is defined as the addition and subsequent removal of the modification by O-GlcNAc Transferase (OGT) and O-GlcNAcase (OGA) provides a novel method for cells to regulate various aspects of gene expression, including RNA polymerase function, epigenetic dynamics, and transcription factor activity. We will focus on the complex relationship between phosphorylation and O-GlcNAcylation in the regulation of the RNA Polymerase II (RNAP II) pre-initiation complex and the regulation of the carboxyl-terminal domain of RNAP II via the synchronous actions of OGT, OGA, and kinases. Additionally, we discuss how O-GlcNAcylation of TATA-box binding protein (TBP) alters cellular metabolism. Next, in a non-exhaustive manner, we will discuss the current literature on how O-GlcNAcylation drives gene transcription in cancer through changes in transcription factor or chromatin remodeling complex functions. We conclude with a discussion of the challenges associated with studying O-GlcNAcylation and present several new approaches for studying O-GlcNAc regulated transcription that will advance our understanding of the role of O-GlcNAc in cancer.

Keywords: transcription; O-GlcNAc; O-GlcNAc transferase; O-GlcNAcase

Citation: Parker, M.P.; Peterson, K.R.; Slawson, C. O-GlcNAcylation and O-GlcNAc Cycling Regulate Gene Transcription: Emerging Roles in Cancer. *Cancers* **2021**, *13*, 1666. https://doi.org/10.3390/cancers13071666

Academic Editors: Ikram El Yazidi-Belkoura and Tony Lefebvre

Received: 19 March 2021
Accepted: 26 March 2021
Published: 1 April 2021

1. O-GlcNAcylation Is a Post-Translational Modification (PTM) That Has Regulatory Roles in Gene Transcription

Cells precisely control gene expression in response to their metabolic state, the availability of building blocks and fuel, and environmental cues [1]. Eukaryotic gene transcription is controlled by many proteins, including the basal transcription machinery, epigenetic chromatin remodeling complexes, and transcription cofactors [2]. O-linked β-N-acetylglucosamine (O-GlcNAc) has been found on proteins in all these groups and is involved with virtually every step of transcription [1]. O-GlcNAc is a PTM in which a single O-GlcNAc moiety is attached to serine and threonine residues of cytoplasmic, nuclear, and mitochondrial proteins [3–5]. Much progress has been made in our understanding of the biochemical, molecular, and physiological effects of O-GlcNAcylation on gene transcription over the past decade due, in part, to the development of improved genetic and pharmacological tools to study its function, which has rapidly accelerated interest in this PTM and its involvement in gene regulation.

O-GlcNAc cycling describes the rapid addition and removal of the sugar from intracellular proteins, which affects their interactions, stability, localization, and activity. O-GlcNAc is covalently attached to substrate proteins by a single enzyme, *O*-linked *N*-acetylglucosamine transferase (OGT; also known as O-GlcNAc transferase), and removed by a single enzyme, *N*-acetyl-β-glucosaminidase (OGA; also known as O-GlcNAase) [6]. There are, however, different isoforms of OGT and OGA, the functional significance of which are beyond the scope of this review. Both enzymes are indispensable for cellular growth, development, and in most cases, survival [5,7–9]. In rapidly dividing tissue, the loss of OGT or OGA is lethal, making both enzymes attractive cancer therapy targets [7].

PTMs are often regulated and/or exert their effect through an assortment of proteins known as "writers", "readers", and "erasers". Simplistically, "writers" add the PTM to substate proteins, whereas "erasers" remove the modification. For example, kinases add phosphate groups to target proteins, and phosphatases remove the phosphate group from proteins. "Readers" recognize the PTM via "reader domains" and integrate the molecular and chemical information of the PTM to drive cellular processes by recruiting effector proteins and enzymes. For example, proteins with SRC homology 2 (SH2) domains recognize and bind phosphate PTMs [10]. These "writer", "reader", and "eraser" functions can exist in a single protein or in multi-protein complexes to control essential cellular processes. The addition and removal of O-GlcNAc marks, which are found on thousands of proteins, is accomplished by a single pair of enzymes, one "writer" and one "eraser". Until recently, there has been little evidence for proteins with an O-GlcNAc "reader domain" to facilitate interaction with O-GlcNAcylated proteins. Toleman et al. identified several human protein candidates, including 14-3-3 isoforms, that bind O-GlcNAc directly and selectively [11]. Despite nearly four decades of work in the field, outside this example, little experimental evidence exists for O-GlcNAc "reader domains". Thus, the existence of O-GlcNAc "reader domains" remains controversial and requires further exploration.

In the absence of O-GlcNAc "reader" proteins, other possibilities regarding the function of protein O-GlcNAcylation must be considered. There is considerable evidence to support the notion that O-GlcNAcylation may sterically promote or prevent protein-protein interactions. This mechanism is suggested by the wide range of biochemical changes O-GlcNAcylation induces in its protein targets, including inhibition or activation of activity, conformational changes, re-localization, and destruction [5,8,9,12–14]. Evidence also indicates that O-GlcNAcylation engages in a complex interplay with other PTMs, most notably, phosphorylation, to potentiate and govern a diverse array of transcriptional and cellular processes. O-GlcNAcylation is almost as common as phosphorylation and is often found in a reciprocal relationship with phosphorylation on the same serine/threonine residues [15]. Experimental data indicate that the interplay between PTMs or the "PTM code" extends beyond the level of single proteins and may coordinate protein complex formation, function, and activity [16]. Thus, the question remains, how does O-GlcNAc manifest itself functionally in gene transcription, and in what way are the transcriptional functions of O-GlcNAc dysregulated in cancer?

2. O-GlcNAcylation Is Sensitive to Metabolite Pools via HBP

Changes in intracellular nutrient metabolite pools directly affect the level uridine-diphosphate *N*-acetyl glucosamine (UDP-GlcNAc), the substrate for OGT and other glycosyltransferases, produced by the HBP [17,18]. The HBP utilizes products from amino acid, fatty acid, nucleotide, and glucose metabolism to generate UDP-GlcNAc. Since multiple metabolites feed into the HBP, UDP-GlcNAc levels are responsive to fluctuations in these metabolic pathways (Figure 1) [19–21]. For example, increasing glucose availability induces a rapid increase in intracellular UDP-GlcNAc levels in a variety of cell types [22–25]. In contrast, glucose depletion results in a reduction of UDP-GlcNAc levels [17]. Similar to glucose, changes in the intracellular glutamine, fatty acid, and nucleotide pools also lead to a modulation of HBP flux and UDP-GlcNAc levels [17]. The prevailing hypothesis in the O-GlcNAc field is that UDP-GlcNAc levels act as a proxy for overall nutrient availability in the cells. Consequently, O-GlcNAcylation can exert pressure on many cellular processes, including gene regulatory mechanisms, in response to changing nutrient and metabolic demands. Interestingly, dysregulated cellular energetics and altered metabolism are now considered a hallmark of all cancers [26]. Every cancer type studied to date has aberrant O-GlcNAc cycling, and a growing number of studies suggest that O-GlcNAcylation constitutes an important regulator of cancer growth and progression [1,27,28]. However, the clinical importance of shifts in cellular O-GlcNAc levels and how these levels are involved with the onset, progression, and metastasis of cancer are still largely unknown [29].

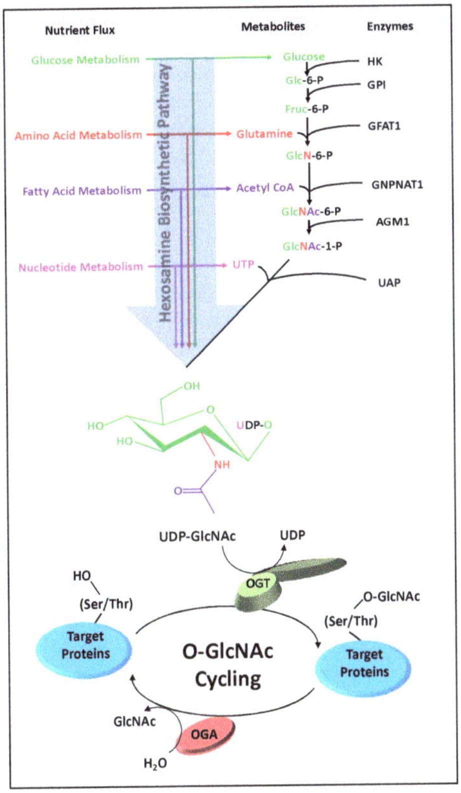

Figure 1. The Hexosamine Biosynthetic Pathway (HBP) and O-GlcNAc Cycling. Glucose enters the cells and is enzymatically converted to Glucose-6-phosphate (Glc-6-P) by Hexokinase (HK). Glucose-6-phosphate isomerase (GPI) then converts Glc-6-P to Fructose-6-phosphate (Fruc-6-P), after which approximately 95% of it proceeds to glycolysis and 3–5% is converted to Glucosamine-6-phosphate (GlcN-6-P) by the enzyme Glutamine fructose-6-phosphate amidotransferase (GFAT). Glutamine is required for this enzymatic reaction. This enzymatic reaction also constitutes the rate-limiting step of the HBP. Glucosamine-6-phosphate *N*-acetyltransferase 1 (GNPNAT1) then utilizes acetyl-CoA to convert GlcN-6-P into *N*-acetylglucosamine-6-phosphate (GlcNAc-6-P). This is then converted to *N*-acetylglucosamine-1-phosphate (GlcNAc-1-P) by Phosphoacetylglucosamine mutase 1 (AGM1). Uridine triphosphate (UTP) is then utilized by UDP-*N*-acetylglucosamine pyrophosphorylase (UAP) to convert GlcNAc-1-P to Uridine diphosphate *N*-acetylglucosamine (UDP-GlcNAc). O-GlcNAc transferase (OGT) and O-GlcNAcase (OGA) facilitate O-GlcNAc cycling "on and off" serine and threonine amino acid residues of target proteins.

3. O-GlcNAcylation Regulates RNA Polymerase Function

At the basal level of transcription, eukaryotic gene expression is governed by three different evolutionarily conserved RNA polymerases (RNAPs). Each RNAP transcribes different types of genes in the genome. RNA polymerase I (RNAP I) transcribes ribosomal RNA genes (rRNAs). RNA polymerase II (RNAP II) transcribes a variety of DNA elements, including messenger RNAs (mRNAs), microRNAs (miRNAs), small nuclear RNAs (snRNAs), and small nucleolar RNAs (snoRNAs). RNA polymerase III (RNAP III) produces transfer RNAs (tRNAs) and 5S ribosomal RNAs (5S rRNAs) [30]. Each RNAP has a dedicated set of core transcription factors that recognize specific conserved promoter elements. These transcription factors and core promoter elements are required to form the pre-initiation complex (PIC) and initiate RNA synthesis.

3.1. *O-GlcNAcylation and RNAP II Function*

Of all the RNAPs and their associated synthesis machinery, eukaryotic RNAP II is the most versatile, as it recognizes and catalyzes transcription from the most diverse set of gene promoters [31]. It also transcribes the widest dynamic range of mRNA expression levels, ranging from just a few mRNA copies for some genes to millions of copies for others [32]. Control of RNAP II activity occurs at many different levels, including recruitment to promoters, PIC formation, transcription initiation, elongation, splicing, and termination. The way in which RNAP II progressively movies through these events is referred to as the transcription cycle. PTMs on the carboxyl-terminal domain (CTD) of RNAP II play a major role in the orchestration of these events. In humans, the CTD of RNAP II consists of 52 imperfect heptad amino acid repeats of predominantly YSPTSPS [33–35]. The phosphorylation status of the CTD defines different forms of RNAP II. RNAP IIA is the unphosphorylated form, and RNAP IIO is the phosphorylated form. Early RNAP II transcriptional cycling models established in the late 1980s and early 1990s suggested that RNAP IIA was the RNAP II spices recruited to gene promoters and required for PIC formation, whereas RNAP IIO was generated later in the transcription cycle during transcriptional initiation [36]. In addition to RNAP IIA and RNAP IIO, a third O-GlcNAcylated species of RNAP II (RNAP IIγ) was identified by Kelly et al. [37]. However, the role of this RNAP species has remained unknown until recently.

Using in vitro cell-free systems derived from crude HeLa cells extracts, Lewis et al. explored the function of RNAP IIγ in the transcription cycle, resulting in a revision of the early model to include O-GlcNAc cycling [15,33,38,39]. In this model, RNAP IIA associates with DNA promoters early in the transcription cycle. RNAP IIA is then O-GlcNAcylated to generate RNAP IIγ during PIC formation. RNAP IIγ is then converted back to the non-O-GlcNAcylated form RNAP IIA by OGA before transcriptional initiation (Figure 2). Both OGT and OGA enzymes activity is essential for PIC formation since inhibition of O-GlcNAc cycling by OGT and OGA inhibitors resulted in transcription inhibition [15,38]. In addition, Lewis et al. found that OGA physically interacts with the elongation factors SPT5 and TIF1β and maps to promoters genome-wide [39]. Human ChIP-seq genomic consortia have shown that O-GlcNAc, OGA, and OGT peaks clearly overlap with the 5′ end of human genes and co-localize with RNAP II peaks. The co-localization of O-GlcNAc, OGA, OGT, and RNAP II at the 5′ end of genes is also observed in *M. musculus*, *Drosophila*, and *C. elegans*, further demonstrating a role for O-GlcNAc cycling in gene transcription that seems to be evolutionally conserved in Eukarya [39,40].

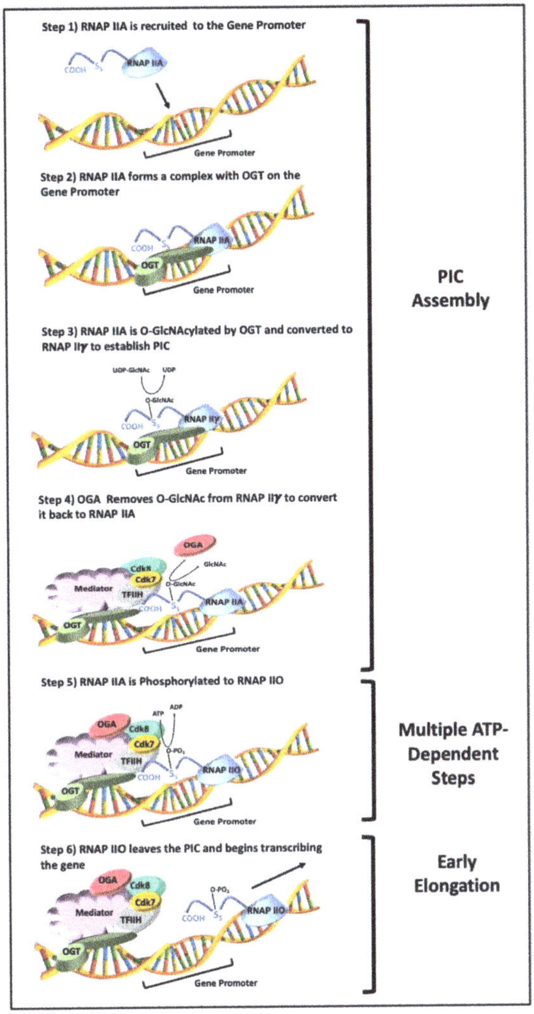

Figure 2. O-GlcNAc cycling and RNAP II-mediated transcription. Model demonstrating how O-GlcNAc cycling regulates various steps of RNAP II transcriptional initiation. In this model, RNAP IIA (unmodified) is recruited to gene promoters by basal transcription factors and core promoter elements (Step 1). RNAP IIA then interacts with OGT (Step 2) and is then O-GlcNAcylated, generating RNAP IIγ (Step 3). OGT may be part of the basal transcription factors at the core promoter elements or is recruited to these *cis*-regulatory elements after RNAP IIA is bound. O-GlcNAcylation of RNAP II, and possibly other basal transcription proteins, lead to the formation of the PIC. OGA associates with the PIC to convert RNAP IIγ back to RNAP IIA (Step 4). Finally, RNAP IIA is phosphorylated to generate RNAP IIO (Step 5). RNAP IIO is released from the PIC to initiate transcription (Step 6).

Functionally, O-GlcNAcylation of the RNAP II CTD may regulate PIC assembly, transcriptional pausing, and elongation in several possible ways. First, O-GlcNAcylation prevents phosphorylation by impeding access of kinases to the necessary serine and threonine residues of the CTD. Phosphorylation of these amino acid residues are required for RNAP II to dissociate from the PIC and critical for mRNA processing and transcription cycle progression. Second, O-GlcNAc addition and removal may promote or impede protein-protein interactions required for PIC assembly, transcriptional pausing, and elongation. Thus, this mechanism may be operative in cells to control which genes are turned off and on, or which gene splice variants are expressed under different internal cellular and external environmental conditions. Beyond RNAP II, mass spectroscopy data have identified 32 additional RNAP II transcription cycling factors that are O-GlcNAcylated [41]. However, little is known about how these modifications affect various aspects of the transcription cycle. Together, these experiments support the notion that there may be a direct connection between the cellular nutrient state, RNAP II function, and transcriptional regulation. In this paradigm of nutrient-regulated gene transcription, the RNAP II transcriptional machinery is constantly sampling the cellular environment via O-GlcNAcylation to govern various aspects of PIC formation, transcription initiation, elongation, splicing, and termination. Based on the data thus far, it is unclear which promoters utilize a GlcNAc-dependent transcription cycling step. To date, only a small number of promoters in cell-free systems have been extensively analyzed. In light of this data, another layer of plasticity must be considered when thinking about the basal transcriptional machinery and how cells might incorporate different environmental and metabolic cues to control gene transcription. In transformed metabolic states such as the Warburg effect and in the microenvironment of solid tumors, these RNAP II nutrient-sensing mechanisms would undoubtedly behave abnormally, leading to altered gene expression and could promote the progression and metastasis of cancer. O-GlcNAcylation may be a mechanism cancer cells use to manipulate their metabolism to promote survival, proliferation, and long-term maintenance in various environmental conditions.

3.2. *O-GlcNAcylation of TATA-Box Binding Protein (TBP) Alters Metabolic Gene Expression*

A protein factor common to all three RNAP complexes is the TBP, which is thought to serve as a PIC assembly platform in TATA-box containing promoters. TBP associates with two RNAP II transcriptional complexes, TFIID and B-TFIID [42,43]. Each complex has unique TBP-associated factors that aid in promoter selectivity. TFIID and B-TFIID are not equivalent; TFIID is responsive to gene-specific transcription factors such as SP1, while B-TFIID is not. Recently, Hardiville et al. identified an O-GlcNAc site in the N-terminal domain (NTD) of TBP at T114 [43]. O-GlcNAcylation at this amino acid residue blocks the ability of TBP to interact with TATA-box binding protein associated factor 1 (BTAF1), which disrupts B-TFIID complex formation. To test the functional significance of O-GlcNAcylation at this amino acid residue, cellular O-GlcNAcylation was increased by inhibiting OGA activity with Thiamet G (TMG). Increasing TBP O-GlcNAcylation had no effect on nuclear localization; however, there was a substantial increase in DNA binding and a significant reduction in BTAF1 interaction. To reduce TBP O-GlcNAcylation, OGT was inhibited with Ac4SGlcNAc, which resulted in decreased TBP chromatin immobilization and increased BTAF1 binding. This data suggests that O-GlcNAc can regulate RNAPs by modulating TBP DNA binding and B-TFIID complex formation. CRISPR/Cas9 mutagenesis of the T114A O-GlcNAc site on TBP increased TBP binding to BTAF1 and directly impacted the expression of 408 genes. This mutation led to profound reprogramming of cellular metabolism and alterations in lipid storage (Figure 3) [43]. This study reveals that the NTD of TBP can integrate nutrient signals via O-GlcNAcylation, to modulate cell metabolism by adjusting gene expression programs.

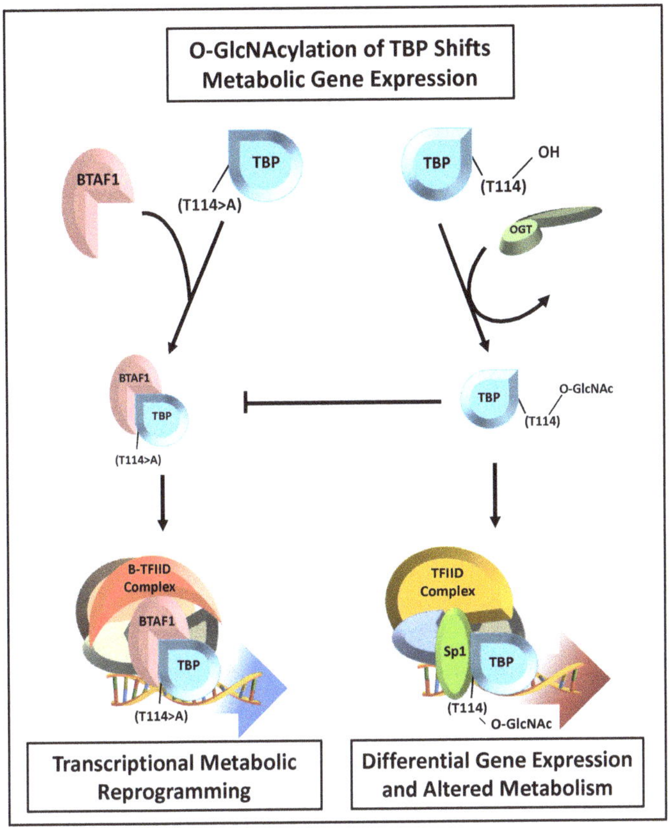

Figure 3. O-GlcNAcylation of TBP regulates metabolic gene expression. O-GlcNAcylation of TBP at the amino acid residue T114 impairs B-TFIID complex formation by disrupting BTAF1 binding. Disruption of TBP O-GlcNAcylation results in significant metabolic transcriptome reprogramming, leading to a profound alteration in lipid storage. Using O-GlcNAc as a nutrient sensor, cells can fine-tune the metabolic transcriptome. Dysregulation of O-GlcNAc signaling has catastrophic effects on metabolic gene expression.

Phosphorylation and O-GlcNAcylation have extensive crosstalk with each other [5,44–46]. Experimental evidence suggests that the simultaneous phosphorylation of the NTD of TBP and transcription factor II B (TFIIB), a general transcription factor involved in the formation of PICs, might stimulate transcription initiation of certain RNAP II regulated genes [47]. However, phosphorylation of TBP alone may result in transcriptional silencing of RNAP II regulated genes during mitosis [48]. In this regard, PTM marks may constitute a complex "code" on the NTD of TBP that fine-tune its activity and promote or obstruct interaction of TBP with specific proteins or protein complexes, thereby controlling the expression of genes regulated by these interactions [43]. In the context of nutrient-regulated gene transcription, this data provides additional evidence of a connection between cell metabolism and transcriptional regulation. Under atypical cellular O-GlcNAc conditions, such as those commonly observed in cancer, the activity of TBP, TFIIB, and genes sensitive to the phosphorylation and O-GlcNAc states of these proteins are at risk for anomalous expression. The metabolic reprogramming mechanism presented by Hardiville et al. [43] may be involved with cancer pathology, but the question remains whether aberrant O-GlcNAcylation is a driver or a consequence of upstream events that led to the manifestation of cancer.

4. O-GlcNAcylation and Epigenetic Gene Regulators

Cancer results from a combination of changes to the genome and epigenome [49]. Epigenetics is defined as the study of changes in gene expression due to mechanisms other than DNA sequence mutations. The alterations in gene expression associated with epigenetics are governed by changes in chromatin structure. Chromatin consists of DNA and its associated proteins, most notably, histones. Epigenetic "marks" are the complete set of DNA and histone modifications that modulate the affinity of chromatin-binding proteins by altering chromatin structure [50–52]. Nutrient flux affects gene expression through epigenetic mechanisms [53]. For example, histone acetyltransferases (HATs) transfer the acetyl group from acetyl-CoA to the lysine residues of histones, thereby neutralizing the positive charge on lysine residues. This reduces the ionic interaction between histones and the DNA sugar-phosphate backbone, which leads to a more relaxed open chromatin conformation. Acetyl-CoA is a metabolite produced by nutrient flux through carbohydrate and lipid metabolic pathways. Elevated levels of acetyl-CoA stimulate cellular growth by prompting histone acetylation of genes controlling growth, which increases the expression of these genes. Chromatin and epigenetic regulatory complexes have been identified as O-GlcNAc targets [54]. O-GlcNAc is also part of the so-called "histone code". Several O-GlcNAc sites have been mapped to histone tails and the histone-DNA interface [53]. Evaluation of histone proteins in HeLa cell nuclear extracts identified O-GlcNAc on histone H2A, H2B, H3, and H4 [55]. Perturbing O-GlcNAc cycling by manipulating OGT and OGA expression leads to alteration of various histone modifications, such as H3K9 acetylation, H3S10 phosphorylation, and H3R17/K27 methylation, strongly implicating O-GlcNAc in epigenetic gene regulatory mechanisms [55,56].

4.1. *O-GlcNAcylation of Ten-Eleven Translocation Protein Family*

Methylation of cytosines in DNA, particularly at promoter CpG islands, is the classical epigenetic modification that plays a critical role in transcription, affecting such downstream processes as chromosome accessibility, nucleosome positioning, and ultimately, gene expression [57]. In humans, three members of the ten-eleven translocation protein family (TET1/2/3) catalyze the sequential oxidation of 5-methylcytosine to 5-hydroxymethylcytosine, 5-formylcytosine, and 5-carboxylacytosine, which serve an essential role in embryonic development and tumor progression [58]. TET proteins interact with OGT and undergo O-GlcNAcylation, which alters their activity and stability [59]. In addition, TET proteins form complexes with OGT and shuttle OGT to specific loci where target proteins are O-GlcNAcylated. For example, Chen et al. have shown that TET2 mediates O-GlcNAcylation of histone H2B at S112 at highly transcribed genes [59]. Deplus et al. showed a direct physical interaction between OGT and TET2/3 proteins and proposed a hierarchical model for H3K4me3 and transcriptional activation [60]. Initially, a TET2 or TET3-OGT complex is formed, which is targeted to DNA loci by an unknown mechanism. OGT does not appear to influence TET2/3 enzymatic activity or targeting; rather, TET2/3 proteins target and promote O-GlcNAcylation of numerous proteins, including host cell factor 1 (HCF1). HCF1 is essential for the recruitment of the SET1/COMPASS complex. O-GlcNAcylation of HCF1 stabilizes the SET1/COMPASS complex and promotes H3K4me3 and subsequent transcriptional activation. Perturbation of TET2/3 or OGT activity results in a direct decrease in H3K4me3 and a concomitant decrease in transcription [60]. More recently, Bauer et al. mapped O-GlcNAc sites on all of the TET proteins [61]. Besides O-GlcNAcylation, TET proteins also were found to be highly phosphorylated, with each TET protein having its own unique phosphorylation pattern or "code" [61]. As O-GlcNAc and phosphorylation are mutually exclusive marks, OGA and, therefore, O-GlcNAc cycling may play an important role in regulating phosphorylation. Collectively, these data show that O-GlcNAcylation is intricately involved with TET1/2/3 protein functions; however, the precise mechanism by which the TET protein-OGT-OGA protein axis works is unclear. The H3K4me3 SET1/COMPASS mechanism proposed by Deplus et al. sheds new light on how TET, OGT, and epigenetic mechanisms work together to regulate gene expression.

Thus, dysregulation of O-GlcNAc in cancer would directly affect the function of TET proteins and their target genes [60].

4.2. O-GlcNAcylation of Polycomb Group Proteins

The polycomb group (PcG) proteins are a diverse family of epigenetic modifiers and transcriptional regulators. In mammals, PcG proteins repress Hox genes and control the expression of other genes that govern embryonic development, X chromosome inactivation, genomic imprinting, the cell cycle, and maintenance of stem cells [62]. Different PcG protein complexes contain distinct chromatin-modifying activities that contribute to the formation of repressive heterochromatin. Dysregulation of PcG proteins plays a role in oncogenesis and is associated with poor patient prognosis [63,64]. Evidence for interaction of OGT with PcG complexes was first observed in Drosophila, where deletion of Ogt led to body plan defects, specifically, abnormal patterning of the anteroposterior axis, a phenotype commonly associated with polycomb gene defects [65]. ChIP-seq experiments performed on Drosophila larvae identified 490 polycomb response elements (PREs) where PcG proteins and OGT co-localized. This data suggested that O-GlcNAcylation is involved with PcG-mediated repression of Hox genes [66–68]. Recently, Geo et al. mapped the OGT interactome in Hela cells and found that OGT interacts with numerous PcG proteins, including those belonging to PRC1, PRC2, PR-DUB, and PhoRC complexes [69]. In an unrelated study, Hauri et al. identified two human PRC2 complexes and two PR-DUB deubiquitylation complexes that contained OGT [70]. Additionally, Forma et al. suggested that EZH2, the catalytic component of the PRC2 complex, interacts with OGT in a cell type-specific manner to regulate a subset of PRC2 target genes such as FOXA1 and FOXC1, two genes that are often dysregulated in hormone-dependent cancers [71]. Interestingly, Jiang et al. found that both OGT and EZH2 are post-transcriptionally inhibited by microRNA-101 (miR-101) [72]. Accumulation of O-GlcNAc, EZH2, and H3K27me3 in the miR-101 promoter region inhibits the transcription of miR-101 and result in upregulation of OGT and EZH2. In colorectal cancer (CRC), elevated protein levels of OGT and EZH2 have been shown to potentiate gene dysregulation and promote CRC metastasis [72]. Thus, manipulation of this regulatory circuit may be a potential therapeutic strategy for metastatic CRC [72]. Decourcelle et al. recently presented additional evidence for OGT and EZH2 involvement in CRC metastasis [73]. They found that O-GlcNAcylated EZH2 transcriptionally repressed UNC5A, a tumor suppressor gene frequently epigenetically downregulated in CRC. This study provides additional evidence for a link between nutrition and O-GlcNAc-mediated epigenetic regulatory mechanisms.

Among the OGT-interacting PcG proteins, several are direct targets of OGT and thus are sensitive to O-GlcNAcylation. Pei-Wen et al. have shown that EZH2 is O-GlcNAcylated in the N-terminal region at S73, S76, S84, and S87, which protects EZH2 from ubiquitin-proteasome degradation [74]. Additionally, Chi-Shuen et al. showed that O-GlcNAcylation of EZH2 at S75 is required for EZH2 protein stability and indirectly facilitates PRC2-mediated gene repression [75]. Moreover, O-GlcNAcylation in the catalytic domain at S729 was found to be essential for EZH2 methyltransferase activity, indicating that O-GlcNAc has a direct effect on EZH2 enzymatic function [74]. Lastly, Maury et al. highlighted the importance of O-GlcNAc on RING1B, which is the catalytic subunit of PRC1 complexes [76]. During human embryonic stem cell (hESC) differentiation, O-GlcNAcylation of RING1B decreases at T250/S251. ChIP-seq results show that non-O-GlcNAcylated RING1B is enriched near cell cycle genes, whereas O-GlcNAcylated RING1B co-localizes to neuronal genes [76]. This suggests that O-GlcNAcylation is a mechanism that targets PRC1 complexes to specific loci.

The role of O-GlcNAc cycling in PcG complex formation and function has recently been evaluated in Drosophila. Using RNAi to knockdown Ogt and Oga, Akan et al. reported that PcG-mediated repression was strikingly insensitive to Oga RNAi, in contrast to Ogt RNAi, which suggested that the addition of O-GlcNAc, and not the removal, is essential for PcG complex formation and function [67]. There are several hypotheses that may account for this observation. First, the O-GlcNAc moieties in PcG complexes may be inaccessible to OGA, and therefore, insensitive to Oga perturbation. Second, the level of RNAi knockdown is not sufficient to produce a phenotype in PcG complexes. Third, Oga and O-GlcNAc cycling is not required for PcG complex formation and function. The overall conclusion of all the data summarized in this section demonstrates that O-GlcNAcylation regulates PcG complexes at many different levels and that there is an intricate relationship between O-GlcNAc and PcG proteins. Further work is required to clarify the roles of O-GlcNAcylation in PcG-mediated gene regulation, especially during tumorigenesis and cancer progression.

4.3. O-GlcNAcylation of GATA1 Target Genes and the Sin3A Corepressor Complex

O-GlcNAc cycling exerts effects on gene transcription via protein-protein interaction with co-repressor and co-activator complexes and lineage-specific master transcription factors that function as adapter proteins to shuttle OGT and OGA to specific loci. An example of this is the Sin3A co-repressor complex, and more recently, erythroid-specific GATA-1 complexes [77,78]. Zhang et al. found that GATA-1, an essential master regulator of erythropoiesis, formed a complex with OGT and OGA when erythroid differentiation was induced [77]. When OGA function was disrupted with TMG, red blood cell maturation was impaired. Zhang et al. hypothesized that GATA-1 facilitated O-GlcNAc cycling at specific loci by delivering OGT and OGA to gene promoters and argued that cells might employ this gene expression regulatory mechanism to drive cell differentiation programs [77]. Little is known about this newly discovered mechanism, and as such, more research is needed to explore its relevance in cellular differentiation and its potential function in hematologic malignancies.

Sin3A is a core component of several transcriptional co-repressor complexes [79]. One of these complexes is the Sin3A-histone deacetylase (Sin3A/HDAC) containing complex. Sin3A physically interacts with, and is O-GlcNAcylated by, OGT [78,80]. How Sin3A/HDAC and OGT activities work to silence genes is not fully understood. O-GlcNAcylation and Sin3A/HDAC activity are thought to work synergistically to regulate transcription factor binding, activator/repressor complex formation and activity, and RNAP II function. Yang et al. has proposed a model for how this complex may regulate gene expression [78]. In this model, sequence-specific repressors bind to gene promoters and recruit the OGT containing Sin3A/HDAC complex. O-GlcNAcylation of these repressors could facilitate and stabilize protein-protein interactions critical for gene repression. Alternatively, or simultaneously, O-GlcNAcylation of activator complexes at gene promoters may disrupt essential hydrophobic interactions and trigger disassembly of these activator complexes, thereby increasing Sin3A/HDAC accessibility to histone targets. OGT activity at the promoter could also potentially arrest RNAP II activity by blocking phosphorylation sites, thereby preventing downstream transcription cycle events. Additionally, O-GlcNAcylation of Sin3A might regulate HDAC activity. Overall, the role of OGT in the Sin3A/HDAC complex could ultimately serve to couple nutrient signals to histone deacetylation activity, thereby regulating gene expression to ensure genes are silenced in an efficient, specific, and nutrient-regulated manner. The discovery of OGT containing Sin3A/HDAC complexes and the model proposed by Yang et al. [78], serve as an example of how O-GlcNAc might collaborate with epigenetic regulatory complexes in general. More experimental data is needed to refine our understanding of how OGT containing epigenetic complexes function, and further exploration is required to examine whether these regulatory mechanisms are dysregulated in metabolic diseases and cancer.

4.4. O-GlcNAcylation of Nucleosome Remodeling Deacetylase

The nucleosome remodeling deacetylase (NuRD) complex is a group of associated proteins with both ATP-dependent chromatin remodeling and histone deacetylase activities that relies on O-GlcNAcylation to regulate its function. NuRD was originally defined as a repressor, but more recently, it has been shown to function both as a co-activator and co-repressor [81]. The core protein constituents of NuRD consist of CHD3/4/5, CDK2AP1, GATAD2A/B, and MBD2/3, which bridges the remodeling subcomplex to the histone deacetylase subcomplex. The composition of the histone deacetylase subcomplex consists of HDAC1/2, MTA1/2/3, and RBBP4/7 proteins [82]. Recent studies have identified O-GlcNAc on every protein of the NuRD complex [69,83]. Additionally, Geo et al. [69] and Zhang et al. [83] demonstrated that OGT directly interacts with the NuRD complex. Zhang et al. showed that when K562 cells were treated with TMG, γ-globin gene expression was reduced, and NuRD occupancy at the promoter increased [83]. OGA was found to interact with NuRD under all conditions tested; however, OGT only interacted with NuRD when γ-globin was silenced [83]. They also demonstrated that O-GlcNAcylation of CHD4 stimulated NuRD complex formation and gene repression of the γ-globin locus, whereas removal of O-GlcNAc from CHD4 by OGA had an activating effect on γ-globin gene expression (Figure 4) [83]. This study provides evidence for the existence of a link between O-GlcNAcylation and NuRD complex formation and function.

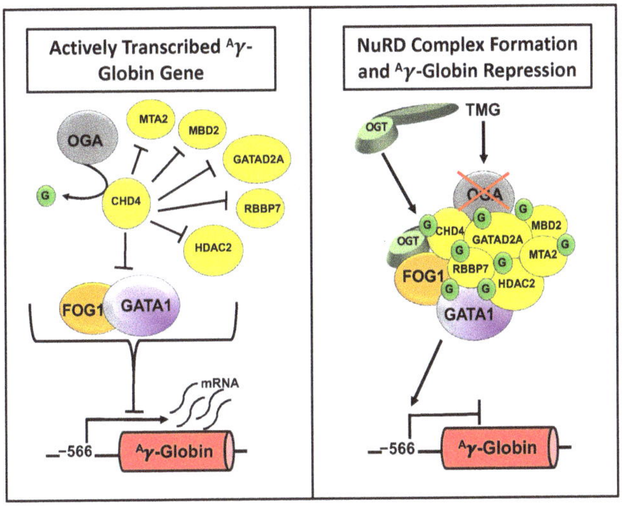

Figure 4. O-GlcNAcylation regulates the formation and function of the NuRD complex. $^{A}\gamma$-globin expression is silenced by the recruitment of NuRD complex to the –566 GATA binding site. OGA removes O-GlcNAc (green circle "G") from CHD4 (a subunit of the NuRD complex), thereby preventing the assembly of the NuRD complex at the –566 site (left panel). When OGA is inhibited by TMG, OGT associates and O-GlcNAcylates CHD4, stimulating the formation of the NuRD complex at the –566 site, which represses $^{A}\gamma$-globin gene transcription (right panel). Additional O-GlcNAc sites have been found on every subunit of the NuRD complex, however, their function remains unknown.

In several cervical cancer cell lines, Geo et al. found that OGT knockdown and overexpression altered Snail gene expression [69]. Snail is a DNA-binding transcriptional repressor known to interact with RNAP II and drive epithelial-mesenchymal transition and metastasis when dysregulated [84,85]. Interestingly, the Snail gene was shown to be regulated by the NuRD complex in breast cancer [86]. Thus, OGT, O-GlcNAc cycling, and NuRD may function coordinately to regulate the expression of the Snail gene and potentially other oncogenic and tumor suppressor genes [69,86]. Similar to the OGT containing Sin3A/HDAC complex, regulation may occur at multiple levels. The NuRD complex plays essential roles in chromatin assembly, transcription, cell cycle progression, and genomic stability [87]. The regulation of the NuRD complex by O-GlcNAc cycling represents a largely understudied area of epigenetic gene regulation. Decades of research demonstrate that cancer is not only a gene-based disease but can also arise from epigenetic abnormalities [88]. Aberrant epigenetic gene regulation is a major contributing factor to tumorigenesis, metastasis, and chemotherapy resistance [89]. Thus, understanding O-GlcNAc-mediated NuRD complex formation and function will undoubtedly lead to a better understanding of cancer pathology and potentially other human diseases.

5. Defects in O-GlcNAcylation of Transcription Factors Promote Cancer

Almost all RNAP II-associated transcription factors are O-GlcNAcylated, often at multiple sites [53]. O-GlcNAc has been found on RNAP II, RNAP accessory transcription factors, co-activators, co-repressors, and lineage-specific transcription factors. O-GlcNAcylation of these transcriptional proteins affects function in various ways, but overall, the major effects are on their activity, localization, and/or stability. Some transcriptional proteins rely solely on one of these properties, for example, activity. Others use all three mechanisms to regulate their function and influence gene transcription. A growing number of studies highlight the importance of O-GlcNAc in transcription factor function and demonstrate how abnormal O-GlcNAcylation, a common phenotype found in all cancers, leads to abnormal gene expression that favors cancer occurrence, progression, and metastasis. In a non-exhaustive manner, examples of recent studies highlighting the role of O-GlcNAcylation of transcription factors and the co-activator yes-associated protein 1 (YAP) in promoting cancer are discussed below.

5.1. O-GlcNAcylation of Transcription Factor Sp1

Specificity protein 1 (Sp1) is a transcription factor that is frequently overexpressed in a wide variety of human cancers and contributes to malignant transformation [90]. Sp1 is a ubiquitous and multifunctional transcription factor that targets promoters that lack a TATA box and are GC rich [91]. Sp1 activates the expression of genes that play a role in tumorigenesis via alterations in angiogenesis, cell growth, differentiation, apoptosis, cellular reprogramming, and heat shock protein gene expression following stress [92,93]. Sp1 is a transcription factor that illustrates the role of O-GlcNAc and O-GlcNAc cycling in transcription [94]. For example, O-GlcNAcylation of Sp1 protects the protein from proteasomal degradation and promotes nuclear localization [91]. Once in the nucleus, O-GlcNAc is removed by OGA so that Sp1 can be phosphorylated, which promotes DNA binding [95]. Thus, O-GlcNAcylation of Sp1 appears to play a central role in its nuclear localization, transactivation, and stability (Figure 5).

5.2. O-GlcNAcylation of Pluripotent Transcription Factors Sox2 and Oct4

Recent studies have shown that Sox2 and Oct4, two transcription factors critical for the induction and maintenance of pluripotency, are O-GlcNAcylated. In many cancers, Sox2 and Oct4 are inappropriately activated, leading to aberrant expression of downstream target genes, which stimulates tumor growth and tumor recurrence. In mouse embryonic stem cells, Olivier-Van Stichelen et al. found that Oga gene ablation resulted in a significant elevation of Sox2 mRNA transcripts [96]. This suggests that increased cellular O-GlcNAcylation favors the expression of Sox2. Studies of hepatocellular carcinomas support this hypothesis. Cao et al. used RNA-ChIP assays to show that eukaryotic initiation factor 4E (eIF4E), a protein stabilized by O-GlcNAc, was strongly associated with the 5′ G/C rich UTR of the Sox2 transcript, thereby enhancing its translation [97]. At the protein level, Sharma et al. established that O-GlcNAcylation of Sox2 increased its transcriptional activity by enhancing protein stability and nuclear localization [98]. These data suggest that O-GlcNAcylation positively regulates Sox2 at multiple leaves. Oct4 is another key pluripotency transcription factor that often co-regulates genes with Sox2. Constable et al. found that the human Oct4 protein is extensively modified by O-GlcNAc [99]. They identified several novel O-GlcNAc sites that might play a role in controlling Oct4 promoter selectivity. In addition, independent of catalytic function, OGT was suggested to function as a bridge protein between Oct4 and Sox2 to activate transcription at specific promoters. In the hyper-O-GlcNAcylated intracellular environment found in many cancers, these two transcription factors may inappropriately activate pluripotency genes, thereby promoting cancer growth and reoccurrence.

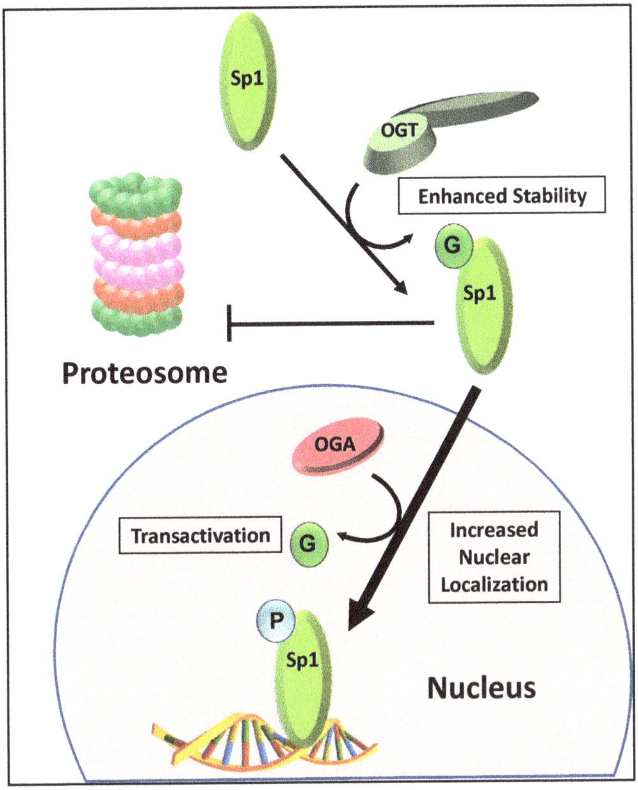

Figure 5. O-GlcNAcylation regulates the stability, nuclear localization, and transactivation of Sp1. Sp1 is a transcription factor that serves as a prototype of how transcription factors are regulated by O-GlcNAcylation at multiple levels. The O-GlcNAcylation of Sp1 (green circle "G") prevents proteasomal degradation and promotes nuclear localization (large arrow). Once in the nucleus, OGA must remove O-GlcNAc from Sp1 so that it can be phosphorylated to facilitate binding to DNA.

5.3. O-GlcNAcylation of Transcription Factors in Breast Cancer

Several studies show that increased O-GlcNAcylation is a key driver of primary and metastatic breast cancer [100–106]. Approximately 70% of all breast cancers express estrogen receptor (ER) and progesterone receptor (PR). O-GlcNAcylation of ER increases the stability of ER by preventing phosphorylation at O-GlcNAcylated amino acid residues [107–110]. Interaction between ER and PR enhances ER DNA-binding and target gene expression [111]. Trinca et al. recently found a novel interaction between PR and OGT, leading to O-GlcNAcylation of PR [106]. Elevated levels of O-GlcNAcylation increased PR-mediated transcriptional activity and altered PR-regulated transcriptional networks in breast cancer cells [106]. In addition to ER and PR, Liu et al. reported that the pioneer transcription factor, FOXA1, is modified by O-GlcNAc in several breast cancer cell lines [112]. O-GlcNAcylation of FOXA1 reduces protein stability, leading to the downregulation of the pro-apoptotic Bim protein, thereby inhibiting apoptosis in breast cancer cells [112]. Under the elevated glucose conditions found in breast cancer cells, the expression of key drug resistance proteins increases, all of which are regulated by the Hedgehog pathway. GLI1 and GLI2 are transcription factors that regulate the Hedgehog pathway and were recently found to be modified by OGT [113]. O-GlcNAcylation of these two proteins enhances their transcriptional activity, resulting in activation of the Hedgehog pathway. Collectively, these studies provide explanations as to why elevated O-GlcNAc levels in breast cancer are associated with poor patient prognosis; however, more work is required to probe the intricacies of these gene regulatory mechanisms.

5.4. O-GlcNAcylation Regulates the Hippo Pathway Co-Activator YAP

The Hippo pathway is an important signaling pathway that controls organ size by regulating cell proliferation and apoptosis. Dysregulation of this pathway has been linked to tumorigenesis [114–117]. In mammalian tissues, the major constituents of the Hippo signaling pathway include the transcriptional co-activator yes-associated protein 1 (YAP), nuclear transcription factors (TEAD1/2/3/4), and their upstream kinases (MST1/2 and LATS1/2) [114,118,119]. In response to unfavorable growth conditions, MST1/2 phosphorylates and activates LATS1/2. Activated LATS1/2 then phosphorylates YAP at S127 and/or S381. Phosphorylation of S127 promotes 14-3-3 protein interaction, resulting in cytoplasmic localization of YAP, whereas phosphorylation of S381 leads to YAP degradation by the SCF(β-TrCP) complex [120,121]. Both events inactivate the Hippo pathway [122]. In an activated Hippo pathway, unphosphorylated YAP translocates to the nucleus and functions as a transcriptional co-activator with the TEAD family of transcription factors to stimulate the expression of genes that promote proliferation and impede apoptosis [122–125].

Recently, Peng et al. demonstrated that this pathway is directly regulated by O-GlcNAcylation [114]. YAP, the core component of the Hippo pathway, interacts with OGT and is O-GlcNAcylated at S109. O-GlcNAcylation of YAP promotes nuclear localization and activation of TEAD downstream target genes. In light of these findings, Zhang et al. reported that O-GlcNAcylation induces transformative phenotypes of liver cancer cells in a YAP-dependent manner [126]. They also identified another O-GlcNAc site on YAP at T241. Mutation of this amino acid residue increased YAP phosphorylation, which led to decreased YAP stability and pro-tumorigenic capacity [126]. Interestingly, both studies found that YAP regulates OGT gene expression and that these two proteins enter into a positive feedback loop to drive YAP O-GlcNAcylation [114]. At least for liver cancer, these studies suggest that elevated cellular O-GlcNAc levels promote YAP activation, and consequently, the expression of pro-tumorigenic genes. Elevated cellular O-GlcNAcylation further exacerbates the YAP-OGT positive feedback loop. Further studies are required to determine whether this positive feedback loop exists in other cancer types.

The transcription factors and co-activator described here are a small cross-section representative of transcription factors and co-activators known the be modified by OGT that function in cancer. More work is required to explore how O-GlcNAcylation of transcription factors affects the expression of genes that regulate cancer occurrence, progression, and metastasis. In the context of nutrient-regulated gene transcription, the data suggest that cells have evolved mechanisms to connect the overall metabolic state of the cell to transcription factor function. Under normal physiologic conditions, this ensures that in times of "feast and famine", cells elicit the appropriate transcriptional response, and that uncontrolled O-GlcNAcylation of transcription factors plays a role in the onset of cancer.

6. Challenges Associated with O-GlcNAc Research

O-GlcNAc is a unique driver of cancer since O-GlcNAcylation affects transcription in a multitude of ways on a global level (Figure 6). Under normal physiological conditions, properly controlled dynamic O-GlcNAc cycling serves to connect nutrient availability and metabolic flux to gene expression, allowing cells to respond to environmental demands. The idea of nutrient-regulated gene transcription and its relationship to cancer is an unexplored research area that demands further study to fully understand the underlying causes of cancer. However, several challenges remain. For example, both OGT and OGA are essential for cellular growth, development, and in most cases, survival, making gene knockout studies problematic [127]. Further complicating O-GlcNAc studies is the fact that

OGT and OGA levels are transcriptionally linked [128,129]. Experimental manipulations that alter the protein level of one enzyme led to compensatory changes in the protein level of the other enzyme. Thus, attempts to alter the overall cellular level of O-GlcNAcylation are short-lived since cells will adjust the expression of OGT or OGA to compensate and restore O-GlcNAc homeostasis [129–131]. Another issue in O-GlcNAc research is the overwhelming number of cellular processes sensitive to O-GlcNAc manipulation. To date, more than 5000 O-GlcNAcylated proteins have been identified. Genetic or pharmacological manipulation of OGT and OGA affect all these proteins making it almost impossible to look at specific O-GlcNAc events. Additionally, because the modification is found on serine and threonine residues, it has extensive crosstalk with phosphorylation. This is problematic, as serine or threonine to alanine substitutions intended to block O-GlcNAcylation also block phosphorylation at these residues. This complicates this experimental approach and can lead to inaccurate data in interpretation.

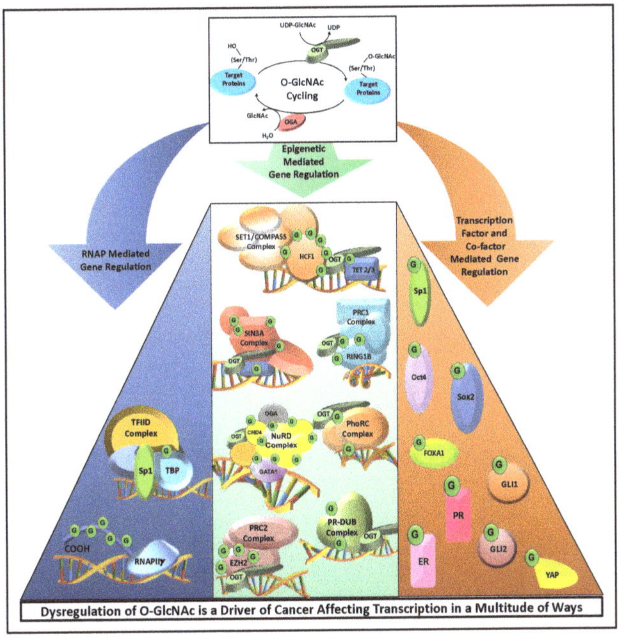

Figure 6. Dysregulation of O-GlcNAc is a driver of cancer by affecting transcription. O-GlcNAcylation is a nutrient-sensitive PTM that modulates gene transcription (green circle "G"). O-GlcNAcylation of the RNAP II CTD affects transcription initiation. It is also found on TBP and numerous RNAP core transcriptional proteins which connects nutrient flux to gene expression. O-GlcNAcylation also influences epigenetic gene regulation via regulation of the SET1/COMPASS, SIN3A, NuRD, PRC1, PRC2, PhoRC, and PR-DUB complexes. Additionally, many transcription factors such as Sp1, Oct4, Sox2, FOXA1, PR, ER, GLI 1, GLI2, and the co-activator YAP all require O-GlcNAcylation to properly regulate their target genes. In general, O-GlcNAcylation of RNAP II, RNAP accessory transcription factors, co-activators, co-repressors, and lineage-specific transcription factors affect their activity, localization, and/or stability. All cancers exhibit aberrant O-GlcNAcylatio; dysregulation of O-GlcNAcylation triggers or exacerbates cancer phenotypes by disrupting normal gene regulation.

The technical challenges for studying dynamic O-GlcNAc cycling remain a major obstacle in the field, so there is a need to develop innovative tools for O-GlcNAc studies that reduce the background noise of global O-GlcNAcylation to allow highly specific, focused experiments that yield unbiased data. Newly developed methodology and technical approaches that selectively modify O-GlcNAcylation on a single protein or hone in on the role of O-GlcNAcylation of proteins at specific DNA loci will be indispensable for elucidating the molecular mechanisms that are governed by the addition or removal of this PTM. Recently, Boulard et al. developed a new CRISPR/Cas9 approach that allows for precise targeting of OGA to any genomic sequence of interest [132]. There is no OGT CRISPR/Cas9 tool available to date. These CRISPR/Cas9 approaches have the potential to precisely target the enzymes to specific DNA sequences, thereby avoiding the pleiotropic effects known to complicate conventional gene knockout, knockdown, and over-expression, as well as pharmacological approaches. Based on the sum total of all data generated thus far, it is evident that O-GlcNAc nutrient-regulated gene expression plays an essential role in many transcriptional processes. Thus, research in this area is essential to understand the molecular mechanisms operative to govern transcriptional processes.

7. Conclusions

There is a rapidly growing interest in how O-GlcNAcylation contributes to the properties of cancer cells and the progression of cancer. Dynamic O-GlcNAc cycling is commonly elevated in cancer cells. Atypical expression and activities of OGT and OGA have been reported in all human cancers studied thus far [133]. An underlying connection between altered cellular metabolism, a major hallmark of cancer, and dynamic O-GlcNAc cycling has been established. As O-GlcNAc cycling has pleiotropic effects within the cell, OGT and OGA are not good direct targets for therapeutic intervention. However, many of the target proteins and molecular regulatory mechanisms that have been, or will be, discovered may be suitable for therapeutic intervention. Understanding the function of this modification in cancer will be enhanced by the advent of new experimental approaches developed to circumvent the technical hurdles inherent in current O-GlcNAc transcriptional studies.

Author Contributions: Conceptualization: M.P.P., K.R.P. and C.S.; resources, K.R.P. and C.S.; writing—original draft preparation: M.P.P.; writing—review and editing, M.P.P., K.R.P. and C.S.; visualization: M.P.P.; supervision: K.R.P. and C.S.; project administration: K.R.P. and C.S.; funding acquisition: M.P.P. and C.S. All authors have read and agreed to the published version of the manuscript.

Funding: This research was funded by National Institute of Health, NIDDK Grant R01DK100595 to K.R.P. and C.S.

Conflicts of Interest: The authors declare no conflict of interest.

References

1. Hardiville, S.; Hart, G.W. Nutrient Regulation of Signaling, Transcription, and Cell Physiology by O-GlcNAcylation. *Cell Metab.* **2014**, *20*, 208–213. [CrossRef]
2. Brimble, S.; Wollaston-Hayden, E.E.; Teo, C.F.; Morris, A.C.; Wells, L. The Role of the O-GlcNAc Modification in Regulating Eukaryotic Gene Expression. *Curr. Signal. Transduct. Ther.* **2010**, *5*, 12–24. [CrossRef]
3. Torres, C.R.; Hart, G.W. Topography and Polypeptide Distribution of Terminal N-acetylglucosamine Residues on the Surfaces of Intact Lymphocytes. Evidence for O-linked GlcNAc. *J. Biol. Chem.* **1984**, *259*, 3308–3317. [CrossRef]
4. Hart, G.W.; Housley, M.P.; Slawson, C. Cycling of O-linked beta-N-acetylglucosamine on Nucleocytoplasmic Proteins. *Nature* **2007**, *446*, 1017–1022. [CrossRef]
5. Hart, G.W.; Slawson, C.; Ramirez-Correa, G.; Lagerlof, O. Cross Talk between O-GlcNAcylation and Phosphorylation: Roles in Signaling, Transcription, and Chronic Disease. *Annu. Rev. Biochem.* **2011**, *80*, 825–858. [CrossRef]
6. Dong, D.L.; Hart, G.W. Purification and Characterization of an O-GlcNAc Selective N-acetyl-beta-D-glucosaminidase from Rat Spleen Cytosol. *J. Biol. Chem.* **1994**, *269*, 19321–19330. [CrossRef]
7. Wang, T.; Birsoy, K.; Hughes, N.W.; Krupczak, K.M.; Post, Y.; Wei, J.J.; Lander, E.S.; Sabatini, D.M. Identification and Characterization of Essential Genes in the Human Genome. *Science* **2015**, *350*, 1096–1101. [CrossRef] [PubMed]
8. Shafi, R.; Iyer, S.P.; Ellies, L.G.; O'Donnell, N.; Marek, K.W.; Chui, D.; Hart, G.W.; Marth, J.D. The O-GlcNAc Transferase Gene Resides on the X Chromosome and is Essential for Embryonic Stem Cell Viability and Mouse Ontogeny. *Proc. Natl. Acad. Sci. USA* **2000**, *97*, 5735–5739. [CrossRef] [PubMed]
9. Keembiyehetty, C.; Love, D.C.; Harwood, K.R.; Gavrilova, O.; Comly, M.E.; Hanover, J.A. Conditional Knock-out Reveals a Requirement for O-linked N-Acetylglucosaminase (O-GlcNAcase) in Metabolic Homeostasis. *J. Biol. Chem.* **2015**, *290*, 7097–7113. [CrossRef] [PubMed]

10. Jin, L.L.; Wybenga-Groot, L.E.; Tong, J.; Taylor, P.; Minden, M.D.; Trudel, S.; McGlade, C.J.; Moran, M.F. Tyrosine Phosphorylation of the Lyn Src Homology 2 (SH2) Domain Modulates Its Binding Affinity and Specificity. *Mol. Cell. Proteom. MCP* **2015**, *14*. [CrossRef] [PubMed]
11. Toleman, C.A.; Schumacher, M.A.; Yu, S.-H.; Zeng, W.; Cox, N.J.; Smith, T.J.; Soderblom, E.J.; Wands, A.M.; Kohler, J.J.; Boyce, M. Structural Basis of O-GlcNAc Recognition by Mammalian 14-3-3 Proteins. *Proc. Natl. Acad. Sci. USA* **2018**, *115*. [CrossRef]
12. Hanover, J.A.; Krause, M.W.; Love, D.C. The Hexosamine Signaling Pathway: O-GlcNAc Cycling in Feast or Famine. *Biochim. Biophys. Acta* **2010**, *1800*. [CrossRef] [PubMed]
13. Bond, M.R.; Hanover, J.A. O-GlcNAc Cycling: A Link between Metabolism and Chronic Disease. *Annu. Rev. Nutr.* **2013**, *33*, 205–229. [CrossRef] [PubMed]
14. Mondoux, M.A.; Love, D.C.; Ghosh, S.K.; Fukushige, T.; Bond, M.; Weerasinghe, G.R.; Hanover, J.A.; Krause, M.W. O-linked-N-acetylglucosamine Cycling and Insulin Signaling are Required for the Glucose Stress Response in Caenorhabditis Elegans. *Genetics* **2011**, *188*. [CrossRef] [PubMed]
15. Lewis, B.A.; Burlingame, A.L.; Myers, S.A. Human RNA Polymerase II Promoter Recruitment in vitro Is Regulated by O-Linked N-Acetylglucosaminyltransferase (OGT). *J. Biol. Chem.* **2016**, *291*, 14056–14061. [CrossRef] [PubMed]
16. Lothrop, A.P.; Torres, M.P.; Fuchs, S.M. Deciphering Post-translational Modification Codes. *FEBS Lett.* **2013**, *587*. [CrossRef]
17. Chiaradonna, F.; Ricciardiello, F.; Palorini, R. The Nutrient-Sensing Hexosamine Biosynthetic Pathway as the Hub of Cancer Metabolic Rewiring. *Cells* **2018**, *7*, 53. [CrossRef] [PubMed]
18. Haltiwanger, R.S.; Holt, G.D.; Hart, G.W. Enzymatic Addition of O-GlcNAc to Nuclear and Cytoplasmic Proteins. Identification of a Uridine Diphospho-N-acetylglucosamine:Peptide Beta-N-acetylglucosaminyltransferase. *J. Biol. Chem.* **1990**, *265*, 2563–2568. [CrossRef]
19. Chaveroux, C.; Sarcinelli, C.; Barbet, V.; Belfeki, S.; Barthelaix, A.; Ferraro-Peyret, C.; Lebecque, S.; Renno, T.; Bruhat, A.; Fafournoux, P.; et al. Nutrient Shortage Triggers the Hexosamine Biosynthetic Pathway via the GCN2-ATF4 Signalling Pathway. *Sci. Rep.* **2016**, *6*, 1–10. [CrossRef]
20. Wang, Z.V.; Deng, Y.; Gao, N.; Pedrozo, Z.; Li, D.L.; Morales, C.R.; Criollo, A.; Luo, X.; Tan, W.; Jiang, N.; et al. Spliced X-box Binding Protein 1 Couples the Unfolded Protein Response to Hexosamine Biosynthetic Pathway. *Cell* **2014**, *156*, 1179–1192. [CrossRef]
21. Ishino, K.; Kudo, M.; Peng, W.X.; Kure, S.; Kawahara, K.; Teduka, K.; Kawamoto, Y.; Kitamura, T.; Fujii, T.; Yamamoto, T.; et al. 2-Deoxy-d-glucose Increases GFAT1 Phosphorylation Resulting in Endoplasmic Reticulum-related Apoptosis via Disruption of Protein N-glycosylation in Pancreatic Cancer Cells. *Biochem. Biophys. Res. Commun.* **2018**, *501*, 668–673. [CrossRef] [PubMed]
22. Marshall, S.; Nadeau, O.; Yamasaki, K. Dynamic Actions of Glucose and Glucosamine on Hexosamine Biosynthesis in Isolated Adipocytes: Differential Effects on Glucosamine 6-phosphate, UDP-N-acetylglucosamine, and ATP Levels. *J. Biol. Chem.* **2004**, *279*. [CrossRef]
23. Schleicher, E.D.; Weigert, C. Role of the Hexosamine Biosynthetic Pathway in Diabetic Nephropathy. *Kidney Int. Suppl.* **2000**, *77*. [CrossRef]
24. Vasconcelos-Dos-Santos, A.; Loponte, H.F.; Mantuano, N.R.; Oliverira, I.A.; De Paula, I.F.; Teixeira, L.K.; De-Freitas-Junior, J.C.; Gondim, K.C.; Heise, N.; Mohana-Borges, R.; et al. Hyperglycemia Exacerbates Colon Cancer Malignancy through Hexosamine Biosynthetic Pathway. *Oncogenesis* **2017**, *6*. [CrossRef] [PubMed]
25. Abdel Rahman, A.M.; Ryczko, M.; Pawling, J.; Dennis, J.W. Probing the Hexosamine Biosynthetic Pathway in Human Tumor Cells by Multitargeted Tandem Mass Spectrometry. *ACS Chem. Biol.* **2013**, *8*. [CrossRef] [PubMed]
26. Akella, N.M.; Ciraku, L.; Reginato, M.J. Fueling the Fire: Emerging Role of the Hexosamine Biosynthetic Pathway in Cancer. *BMC Biol.* **2019**, *17*, 52. [CrossRef]
27. Ma, Z.; Vosseller, K. O-GlcNAc in Cancer Biology. *Amino Acids* **2013**, *45*, 719–733. [CrossRef]
28. Fardini, Y.; Dehennaut, V.; Lefebvre, T.; Issad, T. O-GlcNAcylation: A New Cancer Hallmark? *Front. Endocrinol.* **2013**, *4*, 99. [CrossRef] [PubMed]
29. Slawson, C.; Hart, G.W. O-GlcNAc Signalling: Implications for Cancer Cell Biology. *Nat. Rev. Cancer* **2011**, *11*, 678–684. [CrossRef]
30. Carter, R.; Drouin, G. Structural Differentiation of the Three Eukaryotic RNA Polymerases. *Genomics* **2009**, *94*, 388–396. [CrossRef] [PubMed]
31. Koster, M.J.; Snel, B.; Timmers, H.T. Genesis of Chromatin and Transcription Dynamics in the Origin of Species. *Cell* **2015**, *161*, 724–736. [CrossRef] [PubMed]
32. Levine, M.; Cattoglio, C.; Tjian, R. Looping Back to Leap Forward: Transcription Enters a New Era. *Cell* **2014**, *157*, 13–25. [CrossRef] [PubMed]
33. Lewis, B.A.; Hanover, J.A. O-GlcNAc and the Epigenetic Regulation of Gene Expression. *J. Biol. Chem.* **2014**, *289*, 34440–34448. [CrossRef] [PubMed]
34. Dahmus, M.E. Reversible Phosphorylation of the C-terminal Domain of RNA Polymerase II. *J. Biol. Chem.* **1996**. [CrossRef] [PubMed]
35. Phatnani, H.P.; Greenleaf, A.L. Phosphorylation and Functions of the RNA Polymerase II CTD. *Genes Dev.* **2006**, *20*. [CrossRef] [PubMed]
36. Lu, H.; Flores, O.; Weinmann, R.; Reinberg, D. The Nonphosphorylated form of RNA Polymerase II Preferentially Associates with the Preinitiation Complex. *Proc. Natl. Acad. Sci. USA* **1991**. [CrossRef] [PubMed]

37. Kelly, W.G.; Dahmus, M.E.; Hart, G.W. RNA Polymerase II is a Glycoprotein. Modification of the COOH-terminal Domain by O-GlcNAc. *J. Biol. Chem.* **1993**, *268*, 10416–10424. [CrossRef]
38. Ranuncolo, S.M.; Ghosh, S.; Hanover, J.A.; Hart, G.W.; Lewis, B.A. Evidence of the Involvement of O-GlcNAc-modified Human RNA Polymerase II CTD in Transcription in vitro and in vivo. *J. Biol. Chem.* **2012**, *287*, 23549–23561. [CrossRef] [PubMed]
39. Resto, M.; Kim, B.H.; Fernandez, A.G.; Abraham, B.J.; Zhao, K.; Lewis, B.A. O-GlcNAcase Is an RNA Polymerase II Elongation Factor Coupled to Pausing Factors SPT5 and TIF1β. *J. Biol. Chem.* **2016**, *291*. [CrossRef]
40. Love, D.C.; Ghosh, S.; Mondoux, M.A.; Fukushige, T.; Wang, P.; Wilson, M.A.; Iser, W.B.; Wolkow, C.A.; Krause, M.W.; Hanover, J.A. Dynamic O-GlcNAc Cycling at Promoters of *Caenorhabditis elegans* Genes Regulating Longevity, Stress, and Immunity. *Proc. Natl. Acad. Sci. USA* **2010**. [CrossRef]
41. Lewis, B.A.; Levens, D. O-GlcNAc Transferase Activity is Essential for RNA Pol II Pausing in a Human Cell-Free Transcription System. *bioRxiv* **2020**. [CrossRef]
42. Dynlacht, B.D.; Hoey, T.; Tjian, R. Isolation of Coactivators Associated with the TATA-binding Protein That Mediate Transcriptional Activation. *Cell* **1991**, *66*, 563–576. [CrossRef]
43. Hardiville, S.; Banerjee, P.S.; Selen Alpergin, E.S.; Smith, D.M.; Han, G.; Ma, J.; Talbot, C.C., Jr.; Hu, P.; Wolfgang, M.J.; Hart, G.W. TATA-Box Binding Protein O-GlcNAcylation at T114 Regulates Formation of the B-TFIID Complex and Is Critical for Metabolic Gene Regulation. *Mol. Cell* **2020**, *77*, 1143–1152.e1147. [CrossRef]
44. Leney, A.C.; Atmioui, D.E.; Wu, W.; Ovaa, H.; Heck, A.J.R. Elucidating Crosstalk Mechanisms between Phosphorylation and O-GlcNAcylation. *Proc. Natl. Acad. Sci. USA* **2017**. [CrossRef] [PubMed]
45. Wang, Z.; Gucek, M.; Hart, G.W. Cross-talk between GlcNAcylation and Phosphorylation: Site-specific Phosphorylation Dynamics in Response to Globally Elevated O-GlcNAc. *Proc. Natl. Acad. Sci. USA* **2008**, *105*, 13793–13798. [CrossRef]
46. Bourre, G.; Cantrelle, F.X.; Kamah, A.; Chambraud, B.; Landrieu, I.; Smet-Nocca, C. Direct Crosstalk between O-GlcNAcylation and Phosphorylation of Tau Protein Investigated by NMR Spectroscopy. *Front. Endocrinol.* **2018**, *9*. [CrossRef] [PubMed]
47. Morachis, J.M.; Huang, R.; Emerson, B.M. Identification of Kinase Inhibitors That Target Transcription Initiation by RNA Polymerase II. *Oncotarget* **2011**, *2*. [CrossRef] [PubMed]
48. Segil, N.; Guermah, M.; Hoffmann, A.; Roeder, R.G.; Heintz, N. Mitotic Regulation of TFIID: Inhibition of Activator-dependent Transcription and Changes in Subcellular Localization. *Genes Dev.* **1996**, *10*. [CrossRef] [PubMed]
49. Baylin, S.B.; Jones, P.A. A Decade of Exploring the Cancer Epigenome-Biological and Translational Implications. *Nat. Rev. Cancer* **2011**, *11*, 726–734. [CrossRef] [PubMed]
50. Berger, S.L. The Complex Language of Chromatin Regulation during Transcription. *Nature* **2007**, *447*, 407–412. [CrossRef] [PubMed]
51. Barski, A.; Cuddapah, S.; Cui, K.; Roh, T.-Y.; Schones, D.E.; Wang, Z.; Wei, G.; Chepelev, I.; Zhao, K. High-Resolution Profiling of Histone Methylations in the Human Genome. *Cell* **2007**, *129*, 823–837. [CrossRef] [PubMed]
52. Kouzarides, T. Chromatin Modifications and Their Function. *Cell* **2007**, *128*, 693–705. [CrossRef]
53. Hart, G.W. Nutrient Regulation of Signaling and Transcription. *J. Biol. Chem.* **2019**, *294*, 2211–2231. [CrossRef]
54. Olivier-Van Stichelen, S.; Hanover, J.A. You Are What You Eat: O-linked N-acetylglucosamine in Disease, Development and Epigenetics. *Curr. Opin. Clin. Nutr. Metab. Care* **2015**, *18*. [CrossRef]
55. Sakabe, K.; Wang, Z.; Hart, G.W. β-N-acetylglucosamine (O-GlcNAc) is Part of the Histone Code. *Proc. Natl. Acad. Sci. USA* **2010**. [CrossRef] [PubMed]
56. Sakabe, K.; Hart, G.W. O-GlcNAc Transferase Regulates Mitotic Chromatin Dynamics. *J. Biol. Chem.* **2010**. [CrossRef] [PubMed]
57. Schübeler, D. Function and Information Content of DNA Methylation. *Nature* **2015**, *517*, 321–326. [CrossRef] [PubMed]
58. Li, H.-J.; Wang, Y.; Li, B.-X.; Yang, Y.; Guan, F.; Pang, X.-C. Roles of Ten-eleven Translocation Family Proteins and Their O-linked β-N-acetylglucosaminylated Forms in Cancer Development (Review). *Oncol. Lett.* **2020**, *21*, 1. [CrossRef]
59. Chen, Q.; Chen, Y.; Bian, C.; Fujiki, R.; Yu, X. TET2 Promotes Histone O-GlcNAcylation during Gene Transcription. *Nature* **2013**, *493*, 561–564. [CrossRef] [PubMed]
60. Deplus, R.; Delatte, B.; Schwinn, M.K.; Defrance, M.; Mendez, J.; Murphy, N.; Dawson, M.A.; Volkmar, M.; Putmans, P.; Calonne, E.; et al. TET2 and TET3 Regulate GlcNAcylation and H3K4 Methylation through OGT and SET1/COMPASS. *EMBO J.* **2013**, *32*, 645–655. [CrossRef] [PubMed]
61. Bauer, C.; Gobel, K.; Nagaraj, N.; Colantuoni, C.; Wang, M.; Muller, U.; Kremmer, E.; Rottach, A.; Leonhardt, H. Phosphorylation of TET proteins is regulated via O-GlcNAcylation by the O-linked N-acetylglucosamine transferase (OGT). *J. Biol. Chem.* **2015**, *290*, 4801–4812. [CrossRef]
62. Schuettengruber, B.; Bourbon, H.M.; Di Croce, L.; Cavalli, G. Genome Regulation by Polycomb and Trithorax: 70 Years and Counting. *Cell* **2017**, *171*. [CrossRef] [PubMed]
63. Varambally, S.; Dhanasekaran, S.M.; Zhou, M.; Barrette, T.R.; Kumar-Sinha, C.; Sanda, M.G.; Ghosh, D.; Pienta, K.J.; Sewalt, R.G.; Otte, A.P.; et al. The Polycomb Group Protein EZH2 is Involved in Progression of Prostate Cancer. *Nature* **2002**, *419*. [CrossRef] [PubMed]
64. Benard, A.; Goossens-Beumer, I.J.; Van Hoesel, A.Q.; Horati, H.; Putter, H.; Zeestraten, E.C.; Van De Velde, C.J.; Kuppen, P.J. Prognostic Value of Polycomb Proteins EZH2, BMI1 and SUZ12 and Histone Modification H3K27me3 in Colorectal cancer. *PLoS ONE* **2014**, *9*. [CrossRef] [PubMed]
65. Ingham, P.W. A Gene That Regulates the Bithorax Complex Differentially in Larval and Adult Cells of Drosophila. *Cell* **1984**, *37*, 815–823. [CrossRef]

66. Gambetta, M.C.; Oktaba, K.; Muller, J. Essential Role of the Glycosyltransferase sxc/Ogt in Polycomb Repression. *Science* **2009**, *325*, 93–96. [CrossRef]
67. Akan, I.; Love, D.C.; Harwood, K.R.; Bond, M.R.; Hanover, J.A. Drosophila O-GlcNAcase Deletion Globally Perturbs Chromatin O-GlcNAcylation. *J. Biol. Chem.* **2016**, *291*, 9906–9919. [CrossRef]
68. Liu, T.-W.; Myschyshyn, M.; Sinclair, D.A.; Cecioni, S.; Beja, K.; Honda, B.M.; Morin, R.D.; Vocadlo, D.J. Genome-wide Chemical Mapping of O-GlcNAcylated Proteins in *Drosophila melanogaster*. *Nat. Chem. Biol.* **2016**, *13*, 161–167. [CrossRef] [PubMed]
69. Gao, J.; Yang, Y.; Qiu, R.; Zhang, K.; Teng, X.; Liu, R.; Wang, Y. Proteomic analysis of the OGT Interactome: Novel Links to Epithelial-mesenchymal Transition and Metastasis of Cervical Cancer. *Carcinogenesis* **2018**, *39*, 1222–1234. [CrossRef] [PubMed]
70. Hauri, S.; Comoglio, F.; Seimiya, M.; Gerstung, M.; Glatter, T.; Hansen, K.; Aebersold, R.; Paro, R.; Gstaiger, M.; Beisel, C. A High-Density Map for Navigating the Human Polycomb Complexome. *Cell Rep.* **2016**, *17*. [CrossRef] [PubMed]
71. Forma, E.; Jozwiak, P.; Ciesielski, P.; Zaczek, A.; Starska, K.; Brys, M.; Krzeslak, A. Impact of OGT Deregulation on EZH2 Target Genes FOXA1 and FOXC1 Expression in Breast Cancer Cells. *PLoS ONE* **2018**, *13*. [CrossRef] [PubMed]
72. Jiang, M.; Xu, B.; Li, X.; Shang, Y.; Chu, Y.; Wang, W.; Chen, D.; Wu, N.; Hu, S.; Zhang, S.; et al. Correction: O-GlcNAcylation Promotes Colorectal Cancer Metastasis via the miR-101-O-GlcNAc/EZH2 Regulatory Feedback Circuit. *Oncogene* **2019**, *38*, 5744–5745. [CrossRef] [PubMed]
73. Decourcelle, A.; Very, N.; Djouina, M.; Loison, I.; Thevenet, J.; Body-Malapel, M.; Lelievre, E.; Coqueret, O.; Leprince, D.; El Yazidi-Belkoura, I.; et al. O-GlcNAcylation Links Nutrition to the Epigenetic Downregulation of UNC5A during Colon Carcinogenesis. *Cancers* **2020**, *12*, 3168. [CrossRef] [PubMed]
74. Lo, P.-W.; Shie, J.-J.; Chen, C.-H.; Wu, C.-Y.; Hsu, T.-L.; Wong, C.-H. O-GlcNAcylation Regulates the Stability and Enzymatic Activity of the Histone Methyltransferase EZH2. *Proc. Natl. Acad. Sci. USA* **2018**. [CrossRef] [PubMed]
75. Chu, C.-S.; Lo, P.-W.; Yeh, Y.-H.; Hsu, P.-H.; Peng, S.-H.; Teng, Y.-C.; Kang, M.-L.; Wong, C.-H.; Juan, L.-J. O-GlcNAcylation Regulates EZH2 Protein Stability and Function. *Proc. Natl. Acad. Sci. USA* **2014**. [CrossRef]
76. Maury, J.J.; El Farran, C.A.; Ng, D.; Loh, Y.H.; Bi, X.; Bardor, M.; Choo, A.B. RING1B O-GlcNAcylation Regulates Gene Targeting of Polycomb Repressive Complex 1 in Human Embryonic Stem Cells. *Stem Cell Res.* **2015**, *15*. [CrossRef] [PubMed]
77. Zhang, Z.; Parker, M.P.; Graw, S.; Novikova, L.V.; Fedosyuk, H.; Fontes, J.D.; Koestler, D.C.; Peterson, K.R.; Slawson, C. O-GlcNAc Homeostasis Contributes to Cell Fate Decisions during Hematopoiesis. *J. Biol. Chem.* **2019**, *294*, 1363–1379. [CrossRef]
78. Yang, X.; Zhang, F.; Kudlow, J.E. Recruitment of O-GlcNAc Transferase to Promoters by Corepressor mSin3A: Coupling Protein O-GlcNAcylation to Transcriptional Repression. *Cell* **2002**, *110*, 69–80. [CrossRef]
79. McDonel, P.; Costello, I.; Hendrich, B. Keeping Things Quiet: Roles of NuRD and Sin3 Co-repressor Complexes during Mammalian Development. *Int. J. Biochem. Cell Biol.* **2009**, *41*, 108–116. [CrossRef] [PubMed]
80. Myers, S.A.; Panning, B.; Burlingame, A.L. Polycomb Repressive Complex 2 is Necessary for the Normal Site-specific O-GlcNAc Distribution in Mouse Embryonic Stem Cells. *Proc. Natl. Acad. Sci. USA* **2011**, *108*, 9490–9495. [CrossRef] [PubMed]
81. Bornelov, S.; Reynolds, N.; Xenophontos, M.; Gharbi, S.; Johnstone, E.; Floyd, R.; Ralser, M.; Signolet, J.; Loos, R.; Dietmann, S.; et al. The Nucleosome Remodeling and Deacetylation Complex Modulates Chromatin Structure at Sites of Active Transcription to Fine-Tune Gene Expression. *Mol. Cell* **2018**, *71*. [CrossRef]
82. Hoffmann, A.; Spengler, D. Chromatin Remodeling Complex NuRD in Neurodevelopment and Neurodevelopmental Disorders. *Front. Genet.* **2019**, *10*. [CrossRef] [PubMed]
83. Zhang, Z.; Costa, F.C.; Tan, E.P.; Bushue, N.; DiTacchio, L.; Costello, C.E.; McComb, M.E.; Whelan, S.A.; Peterson, K.R.; Slawson, C. O-Linked N-Acetylglucosamine (O-GlcNAc) Transferase and O-GlcNAcase Interact with Mi2beta Protein at the Agamma-Globin Promoter. *J. Biol. Chem.* **2016**, *291*, 15628–15640. [CrossRef]
84. Cano, A.; Pérez-Moreno, M.A.; Rodrigo, I.; Locascio, A.; Blanco, M.J.; Barrio, M.G.d.; Portillo, F.; Nieto, M.A. The Transcription Factor Snail Controls Epithelial–mesenchymal Transitions by Repressing E-cadherin Expression. *Nat. Cell Biol.* **2000**, *2*, 76–83. [CrossRef] [PubMed]
85. Batlle, E.; Sancho, E.; Franci, C.; Dominguez, D.; Monfar, M.; Baulida, J.; Garcia De Herreros, A. The Transcription Factor Snail is a Repressor of E-cadherin Gene Expression in Epithelial Tumour Cells. *Nat. Cell Biol.* **2000**, *2*. [CrossRef]
86. Fujita, N.; Jaye, D.L.; Kajita, M.; Geigerman, C.; Moreno, C.S.; Wade, P.A. MTA3, a Mi-2/NuRD complex Subunit, Regulates an Invasive Growth Pathway in Breast Cancer. *Cell* **2003**, *113*. [CrossRef]
87. Basta, J.; Rauchman, M. The Nucleosome Remodeling and Deacetylase Complex in Development and Disease. *Transl. Res. J. Lab. Clin. Med.* **2015**, *165*. [CrossRef] [PubMed]
88. Goldberg, A.D.; Allis, C.D.; Bernstein, E. Epigenetics: A Landscape Takes Shape. *Cell* **2007**, *128*. [CrossRef] [PubMed]
89. Sharma, S.; Kelly, T.K.; Jones, P.A. Epigenetics in Cancer. *Carcinogenesis* **2010**, *31*. [CrossRef]
90. Deniaud, E.; Baguet, J.; Mathieu, A.L.; Pages, G.; Marvel, J.; Leverrier, Y. Overexpression of Sp1 Transcription Factor Induces Apoptosis. *Oncogene* **2006**, *25*. [CrossRef]
91. Han, I.; Kudlow, J.E. Reduced O Glycosylation of Sp1 is Associated with Increased Proteasome Susceptibility. *Mol. Cell. Biol.* **1997**, *17*, 2550–2558. [CrossRef]
92. Vellingir, B.; Iyer, M.; Devi Subramaniam, M.; Jayaramayya, K.; Siama, Z.; Giridharan, B.; Narayanasamy, A.; Abdal Dayem, A.; Cho, S.G. Understanding the Role of the Transcription Factor Sp1 in Ovarian Cancer: From Theory to Practice. *Int. J. Mol. Sci.* **2020**, *21*, 1153. [CrossRef]

93. Banerjee, S.; Sangwan, V.; McGinn, O.; Chugh, R.; Dudeja, V.; Vickers, S.M.; Saluja, A.K. Triptolide-induced Cell Death in Pancreatic Cancer is Mediated by O-GlcNAc Modification of Transcription Factor Sp1. *J. Biol. Chem.* **2013**, *288*. [CrossRef] [PubMed]
94. Yang, X.; Qian, K. Protein O-GlcNAcylation: Emerging Mechanisms and Functions. *Nat. Rev. Mol. Cell Biol.* **2017**, *18*, 452–465. [CrossRef]
95. Yang, X.; Su, K.; Roos, M.D.; Chang, Q.; Paterson, A.J.; Kudlow, J.E. O-linkage of N-acetylglucosamine to Sp1 Activation Domain Inhibits Its Transcriptional Capability. *Proc. Natl. Acad. Sci. USA* **2001**, *98*. [CrossRef]
96. Olivier-Van Stichelen, S.; Wang, P.; Comly, M.; Love, D.C.; Hanover, J.A. Nutrient-driven O-linked N-acetylglucosamine (O-GlcNAc) Cycling Impacts Neurodevelopmental Timing and Metabolism. *J. Biol. Chem.* **2017**, *292*, 6076–6085. [CrossRef] [PubMed]
97. Cao, B.; Duan, M.; Xing, Y.; Liu, C.; Yang, F.; Li, Y.; Yang, T.; Wei, Y.; Gao, Q.; Jiang, J. O-GlcNAc Transferase Activates Stem-like Cell Potential in Hepatocarcinoma through O-GlcNAcylation of Eukaryotic Initiation Factor 4E. *J. Cell. Mol. Med.* **2019**, *23*, 2384–2398. [CrossRef] [PubMed]
98. Sharma, N.S.; Gupta, V.K.; Dauer, P.; Kesh, K.; Hadad, R.; Giri, B.; Chandra, A.; Dudeja, V.; Slawson, C.; Banerjee, S.; et al. O-GlcNAc Modification of Sox2 Regulates Self-renewal in Pancreatic Cancer by Promoting Its Stability. *Theranostics* **2019**, *9*, 3410–3424. [CrossRef] [PubMed]
99. Constable, S.; Lim, J.M.; Vaidyanathan, K.; Wells, L. O-GlcNAc Transferase Regulates Transcriptional Activity of Human Oct4. *Glycobiology* **2017**, *27*, 927–937. [CrossRef] [PubMed]
100. Gu, Y.; Mi, W.; Ge, Y.; Liu, H.; Fan, Q.; Han, C.; Yang, J.; Han, F.; Lu, X.; Yu, W. GlcNAcylation Plays an Essential Role in Breast Cancer Metastasis. *Cancer Res.* **2010**, *70*. [CrossRef]
101. Champattanachai, V.; Netsirisawan, P.; Chaiyawat, P.; Phueaouan, T.; Charownwattanasatien, R.; Chokchaichamnankit, D.; Punyarit, P.; Srisomsap, C.; Svasti, J. Proteomic Analysis and Abrogated Expression of O-GlcNAcylated Proteins Associated with Primary Breast Cancer. *Proteomics* **2013**, *13*. [CrossRef] [PubMed]
102. Ferrer, C.M.; Lynch, T.P.; Sodi, V.L.; Falcone, J.N.; Schwab, L.P.; Peacock, D.L.; Vocadlo, D.J.; Seagroves, T.N.; Reginato, M.J. O-GlcNAcylation Regulates Cancer Metabolism and Survival Stress Signaling via Regulation of the HIF-1 Pathway. *Mol. Cell* **2014**, *54*. [CrossRef] [PubMed]
103. Krzeslak, A.; Forma, E.; Bernaciak, M.; Romanowicz, H.; Brys, M. Gene Expression of O-GlcNAc Cycling Enzymes in Human Breast Cancers. *Clin. Exp. Med.* **2012**, *12*. [CrossRef] [PubMed]
104. Caldwell, S.A.; Jackson, S.R.; Shahriari, K.S.; Lynch, T.P.; Sethi, G.; Walker, S.; Vosseller, K.; Reginato, M.J. Nutrient Sensor O-GlcNAc Transferase Regulates Breast Cancer Tumorigenesis through Targeting of the Oncogenic Transcription Factor FoxM1. *Oncogene* **2010**, *29*, 2831–2842. [CrossRef]
105. Sodi, V.L.; Khaku, S.; Krutilina, R.; Schwab, L.P.; Vocadlo, D.J.; Seagroves, T.N.; Reginato, M.J. mTOR/MYC Axis Regulates O-GlcNAc Transferase Expression and O-GlcNAcylation in Breast Cancer. *Mol. Cancer Res.* **2015**, *13*, 923–933. [CrossRef]
106. Trinca, G.M.; Goodman, M.L.; Papachristou, E.K.; D'Santos, C.S.; Chalise, P.; Madan, R.; Slawson, C.; Hagan, C.R. O-GlcNAc-Dependent Regulation of Progesterone Receptor Function in Breast Cancer. *Horm. Cancer* **2018**, *9*, 12–21. [CrossRef] [PubMed]
107. Jiang, M.S.; Hart, G.W. A Subpopulation of Estrogen Receptors are Modified by O-linked N-acetylglucosamine. *J. Biol. Chem.* **1997**, *272*. [CrossRef]
108. Cheng, X.; Hart, G.W. Glycosylation of the Murine Estrogen Receptor-alpha. *J. Steroid Biochem. Mol. Biol.* **2000**, *75*, 147–158. [CrossRef]
109. Cheng, X.; Hart, G.W. Alternative O-glycosylation/O-phosphorylation of Serine-16 in Murine Estrogen Receptor Beta: Post-translational Regulation of Turnover and Transactivation Activity. *J. Biol. Chem.* **2001**, *276*, 10570–10575. [CrossRef]
110. Ozcan, S.; Andrali, S.S.; Cantrell, J.E. Modulation of Transcription Factor Function by O-GlcNAc Modification. *Biochim. Biophys. Acta* **2010**, *1799*. [CrossRef]
111. Carroll, J.S.; Hickey, T.E.; Tarulli, G.A.; Williams, M.; Tilley, W.D. Deciphering the Divergent Roles of Progestogens in Breast Cancer. *Nat. Rev. Cancer* **2016**, *17*, 54–64. [CrossRef] [PubMed]
112. Liu, Y.; Wang, X.; Zhu, T.; Zhang, N.; Wang, L.; Huang, T.; Cao, Y.; Li, W.; Zhang, J. Resistance to Bortezomib in Breast Cancer Cells That Downregulate Bim through FOXA1 O-GlcNAcylation. *J. Cell. Physiol.* **2019**, *234*, 17527–17537. [CrossRef]
113. Das, S.; Bailey, S.K.; Metge, B.J.; Hanna, A.; Hinshaw, D.C.; Mota, M.; Forero-Torres, A.; Chatham, J.C.; Samant, R.S.; Shevde, L.A. O-GlcNAcylation of GLI Transcription Factors in Hyperglycemic Conditions Augments Hedgehog Activity. *Lab. Invest.* **2019**, *99*, 260–270. [CrossRef]
114. Peng, C.; Zhu, Y.; Zhang, W.; Liao, Q.; Chen, Y.; Zhao, X.; Guo, Q.; Shen, P.; Zhen, B.; Qian, X.; et al. Regulation of the Hippo-YAP Pathway by Glucose Sensor O-GlcNAcylation. *Mol. Cell* **2017**, *68*. [CrossRef] [PubMed]
115. Halder, G.; Johnson, R.L. Hippo Signaling: Growth Control and Beyond. *Development* **2011**. [CrossRef]
116. Pan, D. The Hippo Signaling Pathway in Development and Cancer. *Dev. Cell* **2010**, *19*. [CrossRef] [PubMed]
117. Zhao, B.; Li, L.; Lei, Q.; Guan, K.L. The Hippo-YAP Pathway in Organ Size Control and Tumorigenesis: An Updated Version. *Genes Dev.* **2010**, *24*. [CrossRef] [PubMed]
118. Yu, F.X.; Zhao, B.; Guan, K.L. Hippo Pathway in Organ Size Control, Tissue Homeostasis, and Cancer. *Cell* **2015**, *163*. [CrossRef] [PubMed]
119. Zhao, B.; Tumaneng, K.; Guan, K.-L. The Hippo Pathway in Organ Size Control, Tissue Regeneration and Stem Cell Self-renewal. *Nat. Cell Biol.* **2011**, *13*, 877–883. [CrossRef] [PubMed]

120. Zhao, B.; Wei, X.; Li, W.; Udan, R.S.; Yang, Q.; Kin, J.; Xie, J.; Ikenoue, T.; Yu, J.; Li, L.; et al. Inactivation of YAP Oncoprotein by the Hippo Pathway is Involved in Cell Contact Inhibition and Tissue Growth Control. *Genes Dev.* **2007**, *21*. [CrossRef] [PubMed]
121. Zhoa, B.; Li, L.; Tumaneng, K.; Wang, C.Y.; Guan, K.L. A Coordinated Phosphorylation by Lats and CK1 Regulates YAP Stability through SCF(beta-TRCP). *Genes Dev.* **2010**, *24*. [CrossRef]
122. Meng, Z.; Moroishi, T.; Guan, K.L. Mechanisms of Hippo Pathway Regulation. *Genes Dev.* **2016**, *30*. [CrossRef] [PubMed]
123. Lamar, J.M.; Stern, P.; Liu, H.; Schindler, J.W.; Jiang, Z.-G.; Hynes, R.O. The Hippo Pathway Target, YAP, Promotes Metastasis through Its TEAD-interaction Domain. *Proc. Natl. Acad. Sci. USA* **2012**. [CrossRef] [PubMed]
124. Gumbiner, B.M.; Kim, N.-G. The Hippo-YAP Signaling Pathway and Contact Inhibition of Growth. *J. Cell Sci.* **2014**. [CrossRef] [PubMed]
125. Zanconato, F.; Cordenonsi, M.; Piccolo, S. YAP/TAZ at the Roots of Cancer. *Cancer Cell* **2016**, *29*. [CrossRef]
126. Zhang, X.; Qiao, Y.; Wu, Q.; Chen, Y.; Zou, S.; Liu, X.; Zhu, G.; Zhao, Y.; Chen, Y.; Yu, Y.; et al. The Essential Role of YAP O-GlcNAcylation in High-glucose-stimulated Liver Tumorigenesis. *Nat. Commun.* **2017**, *8*, 15280. [CrossRef] [PubMed]
127. O'Donnell, N.; Zachara, N.E.; Hart, G.W.; Marth, J.D. Ogt-dependent X-chromosome-linked Protein Glycosylation is a Requisite Modification in Somatic Cell Function and Embryo Viability. *Mol. Cell. Biol.* **2004**, *24*, 1680–1690. [CrossRef]
128. Slawson, C.; Copeland, R.J.; Hart, G.W. O-GlcNAc Signaling: A Metabolic Link between Diabetes and Cancer? *Trends Biochem. Sci.* **2010**, *35*, 547–555. [CrossRef]
129. Zhang, Z.; Tan, E.P.; VandenHull, N.J.; Peterson, K.R.; Slawson, C. O-GlcNAcase Expression is Sensitive to Changes in O-GlcNAc Homeostasis. *Front. Endocrinol.* **2014**, *5*, 206. [CrossRef] [PubMed]
130. Kazemi, Z.; Chang, H.; Haserodt, S.; McKen, C.; Zachara, N.E. O-linked Beta-N-acetylglucosamine (O-GlcNAc) Regulates Stress-induced Heat Shock Protein Expression in a GSK-3beta-Dependent Manner. *J. Biol. Chem.* **2010**, *285*, 39096–39107. [CrossRef] [PubMed]
131. Slawson, C.; Zachara, N.E.; Vosseller, K.; Cheung, W.D.; Lane, M.D.; Hart, G.W. Perturbations in O-linked Beta-N-acetylglucosamine Protein Modification Cause Severe Defects in Mitotic Progression and Cytokinesis. *J. Biol. Chem.* **2005**, *280*, 32944–32956. [CrossRef] [PubMed]
132. Boulard, M.; Rucli, S.; Edwards, J.R.; Bestor, T.H. Methylation-directed Glycosylation of Chromatin Factors Represses Retrotransposon Promoters. *Proc. Natl. Acad. Sci. USA* **2020**, *117*, 14292–14298. [CrossRef] [PubMed]
133. Forma, E.; Jóźwiak, P.; Bryś, M.; Krześlak, A. The Potential Role of O-GlcNAc Modification in Cancer Epigenetics. *Cell. Mol. Biol. Lett.* **2014**, *19*, 438–460. [CrossRef] [PubMed]

cancers

Review

Role and Function of O-GlcNAcylation in Cancer

Jii Bum Lee [1,2], Kyoung-Ho Pyo [3,*] and Hye Ryun Kim [2,*]

1 Division of Hemato-Oncology, Wonju Severance Christian Hospital, Yonsei University Wonju College of Medicine, Wonju 26426, Korea; jiibumlee@yonsei.ac.kr
2 Division of Medical Oncology, Department of Internal Medicine, Yonsei Cancer Center, Yonsei University College of Medicine, Seoul 06273, Korea
3 Department of Medical Science, Yonsei University College of Medicine, Seoul 06273, Korea
* Correspondence: pkhpsh@gmail.com (K.-H.P.); nobelg@yuhs.ac.kr (H.R.K.); Tel.: +82-2228-0869 (K.-H.P.); +82-2228-8125 (H.R.K.)

Simple Summary: Despite the rapid advancement in immunotherapy and targeted agents, many patients diagnosed with cancer have poor prognosis with dismal overall survival. One of the key hallmarks of cancer is the ability of cancer cells to reprogram their energy metabolism. O-GlcNAcylation is an emerging potential mechanism for cancer cells to induce proliferation and progression of tumor cells and resistance to chemotherapy. This review summarizes the mechanism behind O-GlcNAcylation and discusses the role of O-GlcNAcylation, including its function with receptor tyrosine kinase and chemo-resistance in cancer, and immune response to cancer and as a prognostic factor. Further pre-clinical studies on O-GlcNAcylation are warranted to assess the clinical efficacy of agents targeting O-GlcNAcylation.

Abstract: Cancer cells are able to reprogram their glucose metabolism and retain energy via glycolysis even under aerobic conditions. They activate the hexosamine biosynthetic pathway (HBP), and the complex interplay of O-linked N-acetylglucosaminylation (O-GlcNAcylation) via deprivation of nutrients or increase in cellular stress results in the proliferation, progression, and metastasis of cancer cells. Notably, cancer is one of the emerging diseases associated with O-GlcNAcylation. In this review, we summarize studies that delineate the role of O-GlcNAcylation in cancer, including its modulation in metastasis, function with receptor tyrosine kinases, and resistance to chemotherapeutic agents, such as cisplatin. In addition, we discuss the function of O-GlcNAcylation in eliciting immune responses associated with immune surveillance in the tumor microenvironment. O-GlcNAcylation is increasingly accepted as one of the key players involved in the activation and differentiation of T cells and macrophages. Finally, we discuss the prognostic role of O-GlcNAcylation and potential therapeutic agents such as O-linked β-N-acetylglucosamine-transferase inhibitors, which may help overcome the resistance mechanism associated with the reprogramming of glucose metabolism.

Keywords: O-GlcNAcylation; O-GlcNAc transferase; O-GlcNAcase; cellular stress; cancer; immune surveillance

Citation: Lee, J.B.; Pyo, K.-H.; Kim, H.R. Role and Function of O-GlcNAcylation in Cancer. *Cancers* **2021**, *13*, 5365. https://doi.org/10.3390/cancers13215365

Academic Editor: Constantin N. Baxevanis

Received: 22 September 2021
Accepted: 20 October 2021
Published: 26 October 2021

1. Introduction

One of the pivotal hallmarks of cancer is reprogramming energy metabolism in cancer cells [1]. Normally, cells process glucose under aerobic conditions and favor glycolysis under anaerobic conditions. However, cancer cells differ and reprogram their glucose metabolism to resourcing their energy metabolism mainly to glycolysis [2]. Cancer cells prefer glycolysis instead of oxidative phosphorylation, despite the presence of high oxygen; the effect known as the Warburg effect [3]. Although mitochondrial oxidative phosphorylation produces ATP with approximately 18-fold higher efficiency than aerobic glycolysis, cancer cells manage to compensate energy metabolism by the HBP [4,5]. Thus, approximately 3–5% of glucose is diverted to the HBP, whereas most of the glucose molecules are metabolized through glycolysis [4].

Notably, one of the emerging mechanisms of cancer metabolism behind this complex interplay of glucose metabolism is O-GlcNAcylation, a non-canonical glycosylation that is activated as a response to stimuli such as nutrient deprivation and cellular stress [6]. Since the first discovery of O-GlcNAcylation in 1984, several hypothesis and key concepts have resulted in the understanding of how cellular O-GlcNAcylation is controlled by nutrients and hormones [7]. The key nutrients and metabolic intermediates, such as glucose, amino acid, fatty acid, and nucleotide, are utilized and converted by the HBP to generate the uridine diphosphate GlcNAC (UDP-GlcNAc), a donor substrate for O-GlcNacylation [5]. Subsequently, UDP-GlcNAc transfers O-linked-β-N-acetylglucosamine (O-GlcNAc) to the enzyme O-GlcNAc transferase (OGT), which attaches O-GlcNAc moieties to the serine and/or threonine residues of substrate proteins, including cytoplasmic, nuclear and mitochondrial proteins [8]. This process results in post-translational modification (PTM) of the substrate proteins known as called O-GlcNAcylation [9]. O-GlcNAcase (OGA) reverses the process by catalyzing the hydrolysis done by the OGT [10]. O-GlcNAcylation differs from other PTMs and is strictly controlled primarily by OGT and OGA. This single pair of enzymes recognize hundreds of protein substrates necessary for O-GlcNac homeostasis [9].

O-GlcNAcylation affects many diseases, including diabetes, diabetic nephropathy, and neurodegenerative disease such as Alzheimer disease [11–15]. In recent years, several studies have addressed the role of protein O-GlcNAcylation in various types of cancer, including the impact of O-GlcNAcylation in proliferation, angiogenesis, and metastasis of cancer cells [16,17]. This review summarizes the general mechanism of O-GlcNAcylation, including substrate recognition by OGT and OGA and functions of cellular O-GlcNAcylation. Finally, we discuss the role of O-GlcNAcylation from the perspectives of cancer, including metastasis, receptor tyrosine kinases (RTKs), resistance to chemotherapy, prognostic marker, tumor microenvironment, and the potential targeting of cellular O-GlcNAcylation as cancer therapeutics.

2. O-GlcNAcylation

2.1. Substrate Recognition by OGT and OGA

Contrary to other PTMs that are regulated by diverse enzymes, O-GlcNAcylation is controlled by a single pair of enzymes, OGT and OGA, which recognize hundreds of protein substrates [6,8] (Figure 1). O-GlcNAc, is the master regulator of detained intron (DI) splicing, and regulates gene expression by controlling and splicing the DI in OGT and OGA [18]. Thus, the balance between OGT and OGA level is maintained in an O-GlcNAc-dependent manner. OGT has three isoforms: nucleocytoplasmic (ncOGT), mitochondrial (mOGT), and short OGT (sOGT). These isoforms differ in their subcellular locations and number of amino-terminal tetratricopeptide repeats (TPRs), and thereby differ in length. While ncOGT and sOGT reside in the cytoplasm and nucleus, mOGT is present in the mitochondria. The unique TPR domain length and locations of these isoforms enable targeting various subsets of proteome [19]. Substrates are specific to the isoforms of OGT, and the three isoforms are expressed differently [20,21]. For instance, mOGT is expressed transiently and in much lower than ncOGT, due to its susceptibility to the cellular glucose level [22,23]. The exact mechanisms for substrate recognition by OGT are yet to be clarified; however, the adaptor protein hypothesis and non-specific O-GlcNAcylation are recognized as the plausible mechanisms [24–27].

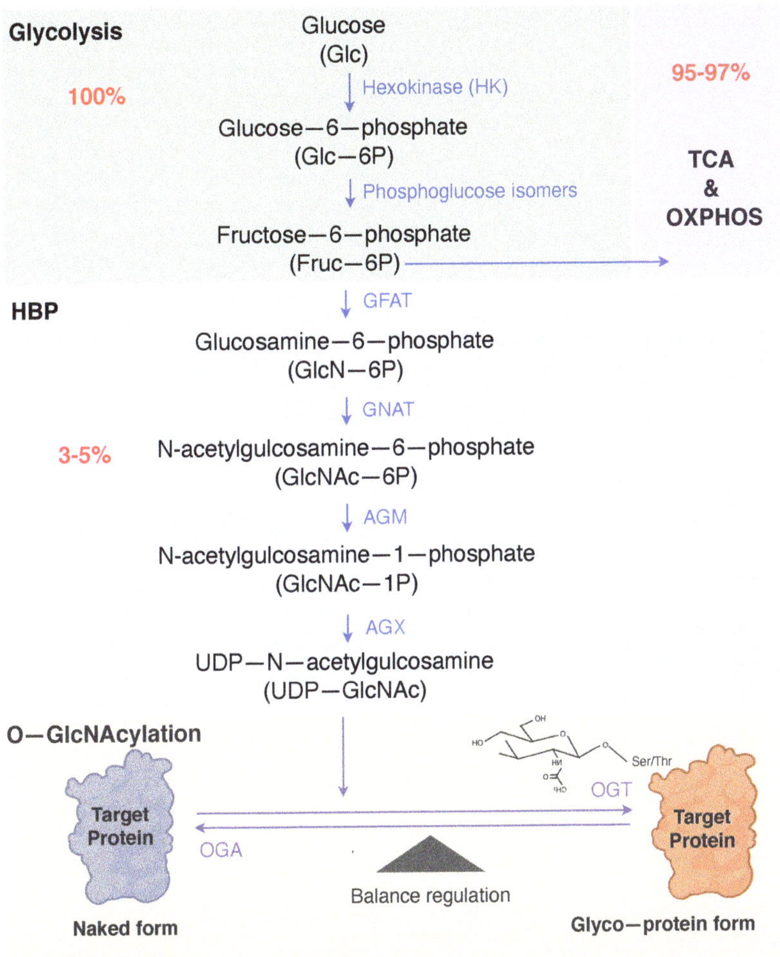

Figure 1. Hexosamine biosynthetic pathway and protein O-GlcNAcylation. Cancer cells compensate for energy metabolism by the hexosamine biosynthetic pathway. While most of the glucose is metabolized through glycolysis, approximately 3–5% of glucose enters the HBP. Glutamine-fructose-6-phosphate aminotransferase, the rate-limiting enzyme of the HBP, converts fructose-6-phosphate into glucosamine-6-phosphate. Subsequently, Glucosamine-6-P is acetylated and uridylation of Glucosamine-1-phosphate generates UDP-N-acetylglucosamine. UDP-GlcNAc acts a substrate, and O-GlcNAc-transferase and O-GlcNAcase add and remove GlcNAc to serine or threonine residues of target proteins, respectively. TCA, Tricarboxylic acid cycle; OXPHOS, Oxidative phosphorylation; HBP, hexosamine biosynthetic pathway GFAT1, hexosamine biosynthetic pathway; GNAT, N-acetyltransferase; AGM, N-acetylphosphoglucosamine mutase; AGX, UDP-N-acetylhexosamine pyrophosphorylase; OGA, O-GlcNAcase; OGT, O-GlcNAc-transferase.

Under specific condition such as glucose deprivation and fasting, OGT substrate recognition may be mediated by proteins, including p38 mitogen-activated protein kinase (MAPK), host cell factor 1 (HCF1), and OGA, which act as adaptor proteins in their receptive substrates of neurofilament H (NFH), peroxisome proliferator-activated receptor-gamma coactivator (PGC)-1alpha (PGC1α), and pyruvate kinase M2 (PKM2), respectively [24–26]. Thus, O-GlcNAcylation is controlled by OGT via adaptor proteins in a context-dependent manner, similar to the ubiquitylation system in which E3 ubiquitin ligases act as adaptor proteins to the E2 ubiquitin-conjugating enzymes [28].

Non-specific O-GlcNAcylation hypothesis stems from the OGT to O-GlcNAcylate proteins in flexible regions, which include loops and termini that bind to the active site, thereby exposing the amide backbone [27]. The preferential modification of substrates that contain the flexible regions by OGA enables the modification of proteins without recognition of any specific sequences or structures. Most mature proteins have a limited number of flexible elements which prevent nonspecific O-GlcNAcylation in normal physiological conditions [29]. During cellular stress, however, non-specific O-GlcNAcylation takes place in unstructured regions of unfolded proteins, facilitating their folding and inhibiting degradation [25,30–32]. For instance, O-GlcNAcylation of unstructured polypeptides, including nascent specificity protein 1 (SP1) and nucleoporin 62 (NUP62) polypeptides, inhibits premature ubiquitin-mediated degradation, and thereby maintains protein homeostasis [33].

The two isoforms of OGA, nucleocytoplasmic isoform and short isoform, differ from each other; with the former having both an N-terminal O-GlcNAc hydrolase domain and C-terminal histone acetyltransferase-like (HAT-like), and the latter lacking the HAT-like domain [13,34]. Currently, information related to the mechanism behind the substrate recognition by OGA is limited due to the incomplete identification of the crystal structure of human OGA [35–37]. Bacterial glycosidases, such as those of *Clostridium perfringenes* and *Bacteroides thetaiotaomicron*, have shown structural similarities by avoiding contact with the peptide chains and instead binding to diverse substrates via interaction of the peptide backbone and sugar moieties [38–40]. Further elucidations of structural features of OGA may identify the mechanism behind the recognition of substrate by OGA.

2.2. Functions of O-GlcNAcylation

Protein O-GlcNAcylation by the complex coordination of OGT and OGA results in regulation of transcription, epigenetic programs, temporal regulation of cell signaling dynamics, and nutrient and hormonal regulation (Figure 2) [41–48]. Both OGT and OGA are O-GlcNAcylated and autoregulated at post-translation level, and cellular stress, such deficient or excess nutrients, disrupts the regulation of cellular O-GcNAc homeostasis [9,11,13,49]. O-GlcNAcylation orchestrates translocation and DNA binding of transcriptional factors such as SP1, RNA polymerase II, and nuclear factor kappa-light-chain-enhancer of activated B cells (NF-κB) in a context-dependent manner [41,43,50]. Proteins such as HCF1 and Ten-eleven translocation (TET), involved in histone modification and DNA methylation, respectively, interact with OGT [34,43,44,51]. OGT regulation is associated with the temporal regulation of insulin signaling dynamics [45]. O-GlcNAcylation negatively regulates insulin signaling when phosphatidylinositol-3,4,5-trisphosphate, the key regulator of insulin signal transduction, recruits OGT from the cytoplasm to the membrane [48,52]. OGT then O-GlcNAcylates and negatively regulates insulin signaling pathway.

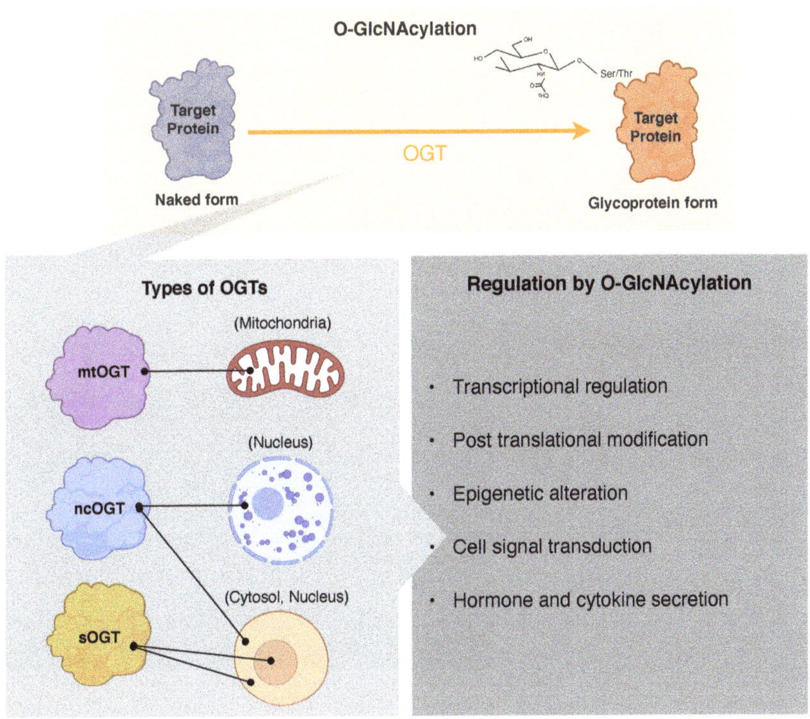

Figure 2. The isoforms of O-GlcNAc transferase (OGT) and the role of O-GlcNAcylation. OGT has three isoforms: nucleocytoplasmic OGT (ncOGT), mitochondrial (mOGT), and short OGT (sOGT). O-GlcNAcylation is involved in transcriptional regulation, post-translational modification, epigenetic alteration, cell signal transduction, and hormone and cytokine secretion.

In addition to the diverse roles O-GlcNAcylation in the fundamental cellular process, its most crucial role is functioning as a nutrient sensor [53]. Previously, increased cellular O-GlcNAcylation was positively associated with the availability of nutrients in response to the influx through the HBP. In vitro and in vivo studies have shown hyperglycemia results in hyper O-GlcNAcylation, thereby supporting the view that HBP flux is the key determinant of an increase in O-GlcNAcylation [46,47]. Recently, evidence has shifted from a positive correlation to variations according to substrate-by-substrate basis [25]. Depending on the levels of nutrients available, OGT and its adaptor proteins modulate the levels of protein O-GlcNAcylation [24–26]. During nutrient deprivation, there is an increase in cellular O-GlcNAcylation, even though UDP-GlcNAc levels are decreased [24,46,47]. The upregulation of OGT expression and the high affinity of OGT for the abundant unfolded proteins during nutrient stress may potentially explain the increase in cellular O-GlcNAcylation despite the decreased level of UDP-GlcNAc. Thus, the complex interplay and harmony of OGT, OGA, and their proteins and substrates result in regulation of cellular protein O-GlcNAcylation.

3. O-GlcNAcylation and Cancer

3.1. O-GlcNAcylation and Metastasis

Several cancers, notably breast, colon, pancreas, liver, and lung cancers, have been associated with elevated O-GlcNAcylation [54,55]. The Cancer Genome Atlas datasets show the aberrant levels of OGT in both adenocarcinoma and squamous cell carcinoma of lung [56]. This results in invasion, metastasis, and angiogenesis of lung cancer cells by activating transcription factors such as Notch receptor 1 (Notch1) and nuclear factor erythroid

2-related factor (Nrf2) [57,58]. Notch-dependent metastasis is potentially modulated via O-GlcNAcylation [57,59].

In NSCLC, O-GlcNAcylation mediates and sustains the epithelial mesenchymal transition (EMT) [60]. EMT markers, such as E-cadherin and vimentin, are suppressed when the O-GlcNAcylation levels are high (Figure 3) [61,62]. Cancer cells that undergo EMT have more aggressive and invasive features due to their ability to migrate [63]. Other transcriptional factors, such as hypoxia-inducible factor(HIF) -1α, nuclear factor kappa-light-chain-enhancer of activated B cells (NF-κB), and signal transducer and activator of transcription 3 (STAT3), are also activated via O-GlcNAcylation, resulting in cancer invasion and metastasis in cancers such as NSCLC, cervical cancer, and head and neck cancer [59,64,65].

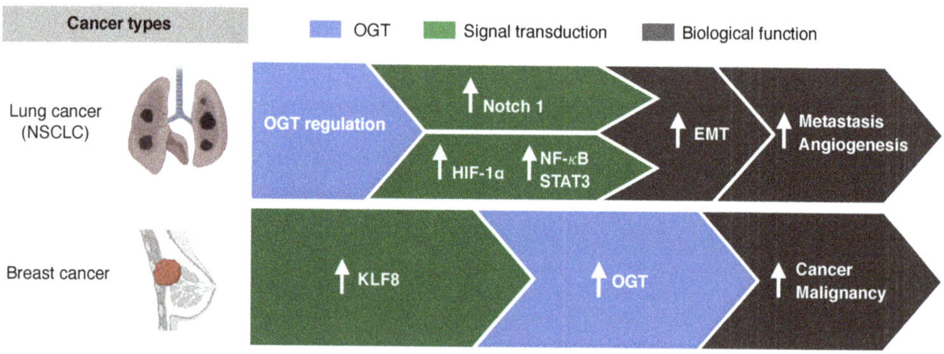

Figure 3. O-GlcNAcylation and metastasis. In non-small cell lung cancer (NSCLC), the epithelial mesenchymal transition (EMT) and transcriptional factors such as hypoxia-inducible factor-1α (HIF) and nuclear factor kappa-light-chain-enhancer of activated B cells (NF-κB) are mediated via O-GlcNAcylation. In breast cancer, the activation of Krüppel-like factor 8 (KLF8) results in OGT activation.

In vitro and in vivo studies on breast cancer cell lines have shown that O-GlcNAc transferase regulates cancer stem-like potential [54]. Krüppel-like factor 8 (KLF8) acts as a novel regulator in mammosphere formation, and activation of KLF8 resulted in activation of OGT in xenograft tumors in vivo. Breast cancers with KLF8 showed worse OS than breast cancers without KLF8 expression. The role of OGT in the regulation of cancer-stemness and tumor metastasis, as seen in breast cancer, may potentially be targeted to overcome resistance to chemotherapy.

3.2. *O-GlcNAcylation and Receptor Tyrosine Kinase*

Activation of downstream pathways of RTK, such as KRAS and epidermal growth factor receptor (EGFR), is associated with an increased glucose flux via the HBP pathway [66,67]. Approximately 90% of the pancreatic ductal adenocarcinoma (PDAC) is associated with KRAS [68]. Increased cellular O-GlcNAcylation initiates reduction in ribonucleotide reductase activity and dNTP pools, resulting in genomic alterations, including *KRAS* mutations in PDAC [66]. Similarly, O-GlcNAcylation is associated with tumorigenesis in *KRAS*-mutant lung cancer [67]. In mouse models, the upregulation and downregulation of O-GlcNAcylation significantly accelerated and delayed $Kras^{G12D}$ lung tumorigenesis, respectively. *EGFR* mutations are also crucial in NSCLC, because more than 60% of patients diagnosed with NSCLC harbor *EGFR* mutations, and treatment with tyrosine kinase inhibitors (TKIs) is preferred over chemotherapy or immunotherapy for these subset of patients [69,70]. Contrary to the cell lines of cervical adenocarcinoma, the cell lines of lung adenocarcinoma are associated with EGFR O-GlcNAcylation in the serine and/or threonine residue(s) [71]. The PTM of EGFR via O-GlcNAcylation may possibly be tumor-specific, and warrants further exploration.

3.3. O-GlcNAcylation and Resistance to Chemotherapy

In lung carcinoma cells, hyper-O-GlcNAcylation is associated with cisplatin resistance via the regulation of either p53 or c-Myc [72]. O-GlcNAcylation of p53 and cMyc results in p53 instability due to ubiquitin-mediated proteasomal degradation and inhibition of c-Myc ubiquitination and degradation, respectively. Recent in vitro and in vivo studies also showed that protein O-GlcNac modification of p53 or c-Myc affects the anti-tumor activity of cisplatin in NSCLC cell lines [73]. Treatment with cisplatin increased the activities of OGT and OGA and decreased the activity of AMP-activated protein kinase. However, inhibition of OGT and OGA by altering O-GlcNAc levels did not result in an increased sensitivity of cisplatin in lung cancer cells.

In many types of cancers, cisplatin is the key chemotherapeutic in both the adjuvant and palliative settings [74]. Identifying the potential mechanism behind cisplatin resistance is an unmet need for patients with cancers, as most of the patients experience disease progression with chemotherapeutic agents [75]. In other cancer cell lines, such as hepatocellular carcinoma, inhibition of O-GlcNAc transferase resulted in enhancement of apoptosis by doxorubicin [76]. Further studies are warranted to elucidate the role of O-GlcNAcylation in the resistance to chemotherapeutic agents used in cancers such as navelbine, pemetrexed, carboplatin, and docetaxel, as well as potential therapeutic agents targeting hyper-O-GlcNAcylation.

3.4. O-GlcNAcylation as Prognostic Marker

Several studies have shown that cancers, including prostate and colorectal cancers, harboring hyper-O-GlcNAcylation are associated with worse prognosis [77,78]. In squamous cell lung cancer, O-GlcNAcylation and increased OGT levels were observed in lung cancer cells compared with the adjacent lung tissue [78]. The expression of OGT in patients with stages II, III, and IV lung adenocarcinomas was higher than that in patients with stage I lung adenocarcinoma [79]. In this study, stage I patients with high OGT expression had shorter recurrence-free survival (RFS) and poor OS. Multivariate analysis revealed that high OGT was a prognostic factor for both RFS and OS, indicating OGT as a potential biomarker in early-stage lung adenocarcinoma. The clinical significance of O-GlcNAcylation is yet to be determined with larger prospective cohorts and validation studies.

4. O-GlcNAcylation and Immune Responses in Cancer

4.1. Overview of Immune System and O-GlcNAcylation

O-GlcNAcylation is highly related to the immune surveillance in the tumor microenvironment (Figure 4). The metabolic shift in immune cells affects T-cell activation and differentiation. Increased amounts of energy metabolites, such as glucose and amino acids, are required for T cells. Specifically, glutamine uptake is essential in activated T cells. Initially, glutamine is converted to a source of oxaloacetate in the tricarboxylic acid cycle (TCA) cycle. The acetyl-CoA is generated by the metabolism of TCA cycle, which allows for greater fatty acid synthesis. The produce of this metabolic pathway serves as a substrate in the HBP [80]. The role and function of O-GlcNAcylation for the two important immune cells constituting the tumor microenvironment have been studied. The effects of O-GlcNAcylation in the tumor microenvironment include dealing with the differentiation and signaling mechanisms of T cells and differentiation and activation of macrophages.

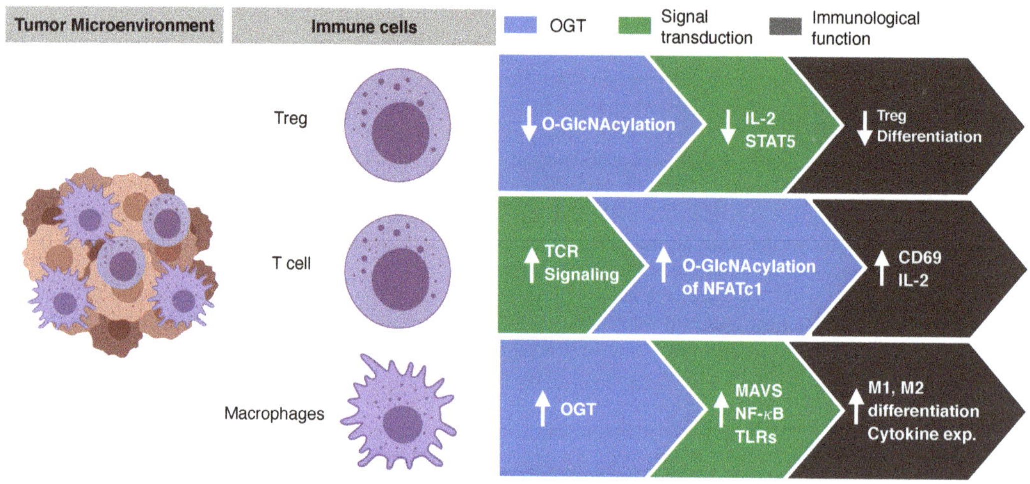

Figure 4. O-GlcNAcylation regulates immune cells activation and differentiation. O-GlcNAcylation is associated with the immune surveillance in tumor microenvironment. O-GlcNAcylation regulates the activation and differentiation of T cells and macrophages, thereby exerting signal transduction and immune surveillance. T reg, regulatory T cell; IL-2, interleukin-2; STAT5, signal transducer and activator of transcription 5; TCR, T-cell receptor; NFATc1, nuclear factor of activated T cells 1; MAVS, mitochondrial antiviral-signaling protein; NF-κB, nuclear factor kappa-light-chain-enhancer of activated B cells; TLRs, Toll-like receptors; exp., expression.

4.2. T Cell Activation and Differentiation Regulated by O-GlcNAcylation

In the thymus, lymphocyte development and activation were observed with increased O-GlcNAc level [81]. This finding suggested that the modification of O-GlcNAc occurred during the early stages of T lymphocyte activation. In addition, the OGT knockdown experiment showed an impairment of interleukin (IL)-2 production in the T cells. Furthermore, ZAP70, SHIP1, and LCK were identified as substrates of O-GlcNAc that regulate TCR signaling [82].

Helper T cells (Th cells) differentiate into several types of cells by the cytokines or surface ligands expressed in antigen-presenting cells or adjacent immune cells. Th cells comprise Th1, Th2, Th17, and regulatory T cells (Treg), which are activated in an abnormal environment, such as an infection or tumor. Finally, differentiated Th in the form of cytokines exhibit anti-tumor and anti-infection effects. However, differentiation of Th can also reduce the anti-tumor effect and damage the host due to excessive immune response. In general, anticancer immune responses are regulated by cellular immunity, and in particular, Th1-type cells activate surrounding cytotoxic T lymphocytes, helping the cytokines kill the tumor cells more effectively. In addition, Th2 and Th17 cells have different roles. Th2 cells induce the activation of B cells through humoral immunity but can reduce the Th1 immune response. Th17 cells promote the differentiation of fibroblast-like cells. Treg cells play an important role in reducing the immune response by depletion of IL-2, the expression of transforming growth factor beta and IL-10, and CTLA-4. Interestingly, it has been reported that O-GlcNAc is involved in the differentiation of Th cells. When TMG, a drug capable of inhibiting OGA, was administered to experimental animals, the levels of O-GlcNAc decreased [83]. Whether Th17 is a friend or foe in terms of tumor immune response is still unclear. In another study, O-GlcNAcylation was found to regulate Foxp3 [84]. Treg is well known as an important regulator of antitumor effect in tumor immunity. In Treg cells deficient in O-GlcNAc, it was confirmed that Foxp3 expression was reduced; and therefore, Treg functions were not effective, thus explaining the regulation of IL-2/STAT5 by O-GlcNAc. A recent study reported that STAT3 and the signal transducer and activator of transcription 5 (STAT5), which play important roles in Th17 cell differentiation and Treg differentiation, respectively, are important factors influencing T cell differentiation and can be regulated by O-GlcNAc [84,85]. In addition, O-GlcNAcylation of key signaling proteins that play an important role in T cells, such as nuclear factor of activated T cells (NFAT), affects the activation and function of T cells [86]. TCR in T cells acts as an important sensor that can detect and kill major histocompatibility complex molecules when presented with a malformed molecule such as cancer. TCR activation rapidly induced O-GlcNAcylation of NFATc1, and O-GlcNAcylated protein was observed in the nucleus within 5 min [86]. These results show that O-GlcNAcylation plays an important role in gene regulation of TF protein. This was verified through the reduction in TCR-induced production of IL-2 and activation markers such as CD69 through the inhibition of OGT [87].

4.3. Macrophage Differentiation and Activation by O-GlcNAcylation

Macrophages are derived from monocytes and play an important role in fighting an infection or the inflammatory response induced by pathogens [88]. In addition, macrophages perform an antigen presentation function through phagocytosis. Macrophages are myeloid-derived cells and are distinct from lymphocytes based on morphology and biological functions [89]. Macrophages in the tumor are divided into M1 and M2 macrophages (majorly M2-type macrophages). M1 macrophages express higher levels of IL-6, IL-10, and tumor necrosis factor alpha, thereby creating an immune suppressive environment. In contrast, M2 macrophages express higher levels of IL-10 and have a greater wound healing effect on the adjacent cells [90]. O-GlcNAcylation has been reported to affect the differentiation of M1 and M2 immune cells [85].

Several articles have reported that a normal level of OGT increases HBP, and that O-GlcNAcylation increases mitochondrial antiviral-signaling proteins (MAVS) to enhance innate immune response. O-GlcNAcylation plays an important role in attenuating infection with vesicular stomach virus [91]. O-GlcNAcylation enhances M1 macrophage polarization and inflammatory immune response [85]. In contrast, it was reported that the activity of HBP plays an important role in M2 macrophage differentiation. The N-glycosylation pathway plays an important role in activating CD206 and CD301, which are important markers of M2 macrophages in metabolic function [92]. Several studies have demonstrated that O-GlcNAcylation is involved in the activation of M1 and M2 macrophages. OGT increases the activity of lipopolysaccharide (LPS)-stimulated NF-κB and the expression of iNOS gene through mSin3a [93]. In addition, it has been confirmed in microglia cells, that c-Rel and p65 are regulated by O-GlcNAcylation. However, it has also been reported that strong O-GlcNAcylation affects the differentiation and activity of macrophages by inhibiting NF-κB p65 signaling [90]. This response is closely related to the response to the Toll-like receptor (TLR), which is an important function of macrophages. TLR4 signaling showed decreased activity by O-GlcNAcylation, revealing that O-GlcNAcylation is an important mechanism to regulate the innate immune response. LPS signaling increase due to lack of OGT is closely related to O-GlcNAcylation of RIPK3, which affects the phosphorylation of NF-κB and ERK. Phosphorylation of RIPK3 is enhanced in the absence of OGT, resulting in increased NF-κB and ERK signals. In addition, the activation of RIPK1, which affects necroptosis, is also associated with O-GlcNAcylation [94]. Activation of NOD2, which plays an important role in the innate immune response, promotes the expression and secretion of cytokines and chemokines. NOD2 is post-translationally modified by O-GlcNAcylation, and its stability and activity is mediated by O-GlcNAcylation [95].

4.4. O-GlcNAcylation and Tumor Microenvironment

Tumor microenvironment (TME) consists of various cells such as cancer cells, vascular endothelial cells, T cells, NK cells, macrophages, fibroblasts, and dendritic cells [96,97]. TME is a fairly complex system, including the metabolic interactions of various cells, and the direct interaction between cancer cells and immune cells. Macrophages and T cells are crucial in the TME [98,99]. Macrophages are involved in the role of tumor-antigen presentation, and immunosuppression in TME. Macrophages differentiate into M1- and

M2-type macrophages, and occupy a high proportion in TME [97]. M1 macrophages are activated by stimuli such as external pathogens and interferon gamma, and have inflammation effects by secreting IL-12, thereby creating a tumor suppressive environment [100]. On the contrary, M2 macrophages are activated by IL-4, IL-10, and IL-13, and are known to have anti-inflammatory and immune suppressive effects on immune cells.

To overcome the immune suppressive TME created by M2 macrophages, various therapeutic agents are being developed to inhibit M2-type macrophage differentiation or increase the ratio of M1-type macrophages [101,102]. OGT may be a promising potential therapeutic target since OGT in macrophages is important in the regulation of phosphorylation of NF-kB, and the expression of iNOS and pro-inflammatory cytokine genes [93]. Further elucidation on the role of O-GlcNAcylation in M1 and M2 will help pave the way for OGT as a therapeutic target.

T cells also play a critical role in tumor suppression, and kill tumors by directly reacting with tumor antigens expressed on the surface of cancer cells (epitope-MHC complex) [103]. O-GlcNAcylation may function in the regulation of T-cell receptor (TCR) and T lymphocyte differentiation [104]. On the other hand, T cells inhibit the differentiation of immune cells, thereby reducing anti-tumor immune responses by OGT-induced T reg [84]. Thus, the dual mechanism of O-GlcNAcylation in T cells requires comprehensive immunological pre-clinical studies to define O-GlcNAcylation and its relationship with the anti-tumor effect in T cells.

5. Cancer Therapeutics Targeting O-GlcNAcylation

The role of hyper-O-GlcNAcylation in metastasis and resistance to chemotherapy in cancer as well as its potential role as a prognostic marker has prompted the development of targets directed at O-GlcNAcylation [17]. Cell lines of breast, colorectal, prostate, and hepatocellular carcinoma treated with investigational OGT inhibitors have shown a significant decrease in tumor growth [105–109]. Recently, tamoxifen-resistant breast cancer cell lines treated with OGT small molecule inhibitor OSMI-1 showed anti-tumor activity via epigenetic activation of the tumor-suppressor ERRFI1 [62]. Similarly, reductions in OGT levels have shown to inhibit growth of lung cancer cells; however, the role of OGT inhibitor in cancers have not been elucidated using an investigational agent [78].

Despite the promising pre-clinical data of OGT inhibitors, many hurdles remain, including the physiologic role of OGT involved in energy metabolism of normal cells [3]. Agents targeting OGT directly, such as small molecules and bisubstrate inhibitors, were initially ideal potential therapeutic options in the treatment of cancers [110]. However, the possibility of off-target toxicities for small molecules targeting OGT, and the inability to permeate through cells may render bisubstrate inhibitors ineffective [111].

6. Conclusions and Future Perspectives

In the past few decades, the mechanism of PTM of proteins has been extensively examined. Among the PTMs, including acetylation, ubiquitylation, and phosphorylation of proteins, the non-canonical glycosylation of O-GlcNAcylation is an emerging mechanism that remains to be fully elucidated. Cancer metabolism via O-GlcNAcylation is relatively unknown, when compared with other cancer hallmarks such as induction of angiogenesis, shift towards genome instability and mutation, sustenance of proliferative signaling, and avoidance of immune destruction.

O-GlcNAcylation plays a crucial role in the complex interplay of glucose metabolism in many diseases. The key concepts of the functions and mechanisms of O-GlcNAcylation in cancer cells have been recently elucidated. So far, we know that O-GlcNAcylation is associated with metastasis, interacts with RTKs, causes resistance to chemotherapy, may serve as a prognostic marker, and plays a role in immune surveillance in the tumor microenvironment. The complexity of O-GlcNAcylation currently hinders the comprehensive understanding of the mechanisms, which allows for proper selection and targeting of agents, such as OGT inhibitors, to exert anti-tumor activities. Further pre-clinical studies on O-GlcNAcylation will pave way for a better understanding of O-GlcNAcylation as a potential therapeutic target in many types of cancer.

Author Contributions: The literature was searched and reviewed by J.B.L. and K.-H.P. The draft of the manuscript was critiqued and validated by H.R.K. All authors have read and agreed to the published version of the manuscript.

Funding: This research was funded by the Bio and Medical Technology Development Program of the National Research Foundation funded by the Ministry of Science and ICT to H.-R.K. (NRF-2019M3A9B6065231, 2017M3A9E9072669) and to J.B.L. (2021R1I1A1A0105744).

Conflicts of Interest: The authors declare no conflict of interest.

References

1. Hanahan, D.; Weinberg, R.A. Hallmarks of Cancer: The Next Generation. *Cell* **2011**, *144*, 646–674. [CrossRef] [PubMed]
2. Fardini, Y.; Dehennaut, V.; Lefebvre, T.; Issad, T. O-GlcNAcylation: A New Cancer Hallmark? *Front. Endocrinol.* **2013**, *4*, 99. [CrossRef] [PubMed]
3. Liberti, M.V.; Locasale, J.W. The Warburg Effect: How Does it Benefit Cancer Cells? *Trends Biochem. Sci.* **2016**, *41*, 211–218. [CrossRef] [PubMed]
4. Marshall, S.; Bacote, V.; Traxinger, R.R. Discovery of a metabolic pathway mediating glucose-induced desensitization of the glucose transport system. Role of hexosamine biosynthesis in the induction of insulin resistance. *J. Biol. Chem.* **1991**, *266*, 4706–4712. [CrossRef]
5. Bond, M.R.; Hanover, J.A. A little sugar goes a long way: The cell biology of O-GlcNAc. *J. Cell Biol.* **2015**, *208*, 869–880. [CrossRef]
6. Hart, G.W.; Slawson, C.; Ramirez-Correa, G.; Lagerlof, O. Cross talk between O-GlcNAcylation and phosphorylation: Roles in signaling, transcription, and chronic disease. *Annu. Rev. Biochem.* **2011**, *80*, 825–858. [CrossRef] [PubMed]
7. Torres, C.R.; Hart, G.W. Topography and polypeptide distribution of terminal N-acetylglucosamine residues on the surfaces of intact lymphocytes. Evidence for O-linked GlcNAc. *J. Biol. Chem.* **1984**, *259*, 3308–3317. [CrossRef]
8. Hart, G.W.; Housley, M.P.; Slawson, C. Cycling of O-linked β-N-acetylglucosamine on nucleocytoplasmic proteins. *Nature* **2007**, *446*, 1017–1022. [CrossRef] [PubMed]
9. Yang, X.; Qian, K. Protein O-GlcNAcylation: Emerging mechanisms and functions. *Nat. Rev. Mol. Cell Biol.* **2017**, *18*, 452–465. [CrossRef] [PubMed]
10. Hu, P.; Shimoji, S.; Hart, G.W. Site-specific interplay between O-GlcNAcylation and phosphorylation in cellular regulation. *FEBS Lett.* **2010**, *584*, 2526–2538. [CrossRef] [PubMed]
11. Yi, W.; Clark, P.M.; Mason, D.E.; Keenan, M.C.; Hill, C.; Goddard, W.A., 3rd; Peters, E.C.; Driggers, E.M.; Hsieh-Wilson, L.C. Phosphofructokinase 1 glycosylation regulates cell growth and metabolism. *Science* **2012**, *337*, 975–980. [CrossRef] [PubMed]
12. Costa, R.; Remigante, A.; Civello, D.A.; Bernardinelli, E.; Szabó, Z.; Morabito, R.; Marino, A.; Sarikas, A.; Patsch, W.; Paulmichl, M.; et al. O-GlcNAcylation Suppresses the Ion Current IClswell by Preventing the Binding of the Protein ICln to α-Integrin. *Front. Cell Dev. Biol.* **2020**, *8*. [CrossRef]
13. Ruan, H.-B.; Singh, J.P.; Li, M.-D.; Wu, J.; Yang, X. Cracking the O-GlcNAc code in metabolism. *Trends Endocrinol. Metab. TEM* **2013**, *24*, 301–309. [CrossRef] [PubMed]
14. Bond, M.R.; Hanover, J.A. O-GlcNAc cycling: A link between metabolism and chronic disease. *Annu. Rev. Nutr.* **2013**, *33*, 205–229. [CrossRef] [PubMed]
15. Zhu, Y.; Shan, X.; Yuzwa, S.A.; Vocadlo, D.J. The Emerging Link between O-GlcNAc and Alzheimer Disease. *J. Biol. Chem.* **2014**, *289*, 34472–34481. [CrossRef]
16. Ferrer, C.M.; Sodi, V.L.; Reginato, M.J. O-GlcNAcylation in Cancer Biology: Linking Metabolism and Signaling. *J. Mol. Biol.* **2016**, *428*, 3282–3294. [CrossRef] [PubMed]
17. Wu, D.; Jin, J.; Qiu, Z.; Liu, D.; Luo, H. Functional Analysis of O-GlcNAcylation in Cancer Metastasis. *Front. Oncol.* **2020**, *10*, 585288. [CrossRef] [PubMed]
18. Tan, Z.-W.; Fei, G.; Paulo, J.A.; Bellaousov, S.; Martin, S.E.S.; Duveau, D.Y.; Thomas, C.J.; Gygi, S.P.; Boutz, P.L.; Walker, S. O-GlcNAc regulates gene expression by controlling detained intron splicing. *Nucleic Acids Res.* **2020**, *48*, 5656–5669. [CrossRef]
19. Love, D.C.; Kochran, J.; Cathey, R.L.; Shin, S.-H.; Hanover, J.A. Mitochondrial and nucleocytoplasmic targeting of O-linked GlcNAc transferase. *J. Cell Sci.* **2003**, *116*, 647–654. [CrossRef] [PubMed]
20. Lazarus, B.D.; Love, D.C.; Hanover, J.A. Recombinant O-GlcNAc transferase isoforms: Identification of O-GlcNAcase, yes tyrosine kinase, and tau as isoform-specific substrates. *Glycobiology* **2006**, *16*, 415–421. [CrossRef] [PubMed]
21. Trapannone, R.; Mariappa, D.; Ferenbach, A.T.; van Aalten, D.M.F. Nucleocytoplasmic human O-GlcNAc transferase is sufficient for O-GlcNAcylation of mitochondrial proteins. *Biochem. J.* **2016**, *473*, 1693–1702. [CrossRef] [PubMed]
22. Sacoman, J.L.; Dagda, R.Y.; Burnham-Marusich, A.R.; Dagda, R.K.; Berninsone, P.M. Mitochondrial O-GlcNAc Transferase (mOGT) Regulates Mitochondrial Structure, Function, and Survival in HeLa Cells. *J. Biol. Chem.* **2017**, *292*, 4499–4518. [CrossRef]

23. Jóźwiak, P.; Ciesielski, P.; Zakrzewski, P.K.; Kozal, K.; Oracz, J.; Budryn, G.; Żyżelewicz, D.; Flament, S.; Vercoutter-Edouart, A.S.; Bray, F.; et al. Mitochondrial O-GlcNAc Transferase Interacts with and Modifies Many Proteins and Its Up-Regulation Affects Mitochondrial Function and Cellular Energy Homeostasis. *Cancers* **2021**, *13*, 2956. [CrossRef] [PubMed]
24. Cheung, W.D.; Hart, G.W. AMP-activated Protein Kinase and p38 MAPK Activate O-GlcNAcylation of Neuronal Proteins during Glucose Deprivation. *J. Biol. Chem.* **2008**, *283*, 13009–13020. [CrossRef] [PubMed]
25. Ruan, H.-B.; Han, X.; Li, M.-D.; Singh, J.P.; Qian, K.; Azarhoush, S.; Zhao, L.; Bennett, A.M.; Samuel, V.T.; Wu, J.; et al. O-GlcNAc transferase/host cell factor C1 complex regulates gluconeogenesis by modulating PGC-1α stability. *Cell Metab.* **2012**, *16*, 226–237. [CrossRef]
26. Whisenhunt, T.R.; Yang, X.; Bowe, D.B.; Paterson, A.J.; Van Tine, B.A.; Kudlow, J.E. Disrupting the enzyme complex regulating O-GlcNAcylation blocks signaling and development. *Glycobiology* **2006**, *16*, 551–563. [CrossRef] [PubMed]
27. Lazarus, M.B.; Nam, Y.; Jiang, J.; Sliz, P.; Walker, S. Structure of human O-GlcNAc transferase and its complex with a peptide substrate. *Nature* **2011**, *469*, 564–567. [CrossRef]
28. Vucic, D.; Dixit, V.M.; Wertz, I.E. Ubiquitylation in apoptosis: A post-translational modification at the edge of life and death. *Nat. Rev. Mol. Cell Biol.* **2011**, *12*, 439–452. [CrossRef] [PubMed]
29. Zachara, N.E.; O'Donnell, N.; Cheung, W.D.; Mercer, J.J.; Marth, J.D.; Hart, G.W. Dynamic O-GlcNAc modification of nucleocytoplasmic proteins in response to stress. A survival response of mammalian cells. *J. Biol. Chem.* **2004**, *279*, 30133–30142. [CrossRef]
30. Ruan, H.B.; Nie, Y.; Yang, X. Regulation of protein degradation by O-GlcNAcylation: Crosstalk with ubiquitination. *Mol. Cell Proteom.* **2013**, *12*, 3489–3497. [CrossRef]
31. Guinez, C.; Losfeld, M.-E.; Cacan, R.; Michalski, J.-C.; Lefebvre, T. Modulation of HSP70 GlcNAc-directed lectin activity by glucose availability and utilization. *Glycobiology* **2005**, *16*, 22–28. [CrossRef] [PubMed]
32. Guinez, C.; Mir, A.-M.; Leroy, Y.; Cacan, R.; Michalski, J.-C.; Lefebvre, T. Hsp70-GlcNAc-binding activity is released by stress, proteasome inhibition, and protein misfolding. *Biochem. Biophys. Res. Commun.* **2007**, *361*, 414–420. [CrossRef] [PubMed]
33. Zhu, Y.; Liu, T.-W.; Cecioni, S.; Eskandari, R.; Zandberg, W.F.; Vocadlo, D.J. O-GlcNAc occurs cotranslationally to stabilize nascent polypeptide chains. *Nat. Chem. Biol.* **2015**, *11*, 319–325. [CrossRef]
34. Hanover, J.A.; Krause, M.W.; Love, D.C. Bittersweet memories: Linking metabolism to epigenetics through O-GlcNAcylation. *Nat. Rev. Mol. Cell Biol.* **2012**, *13*, 312–321. [CrossRef] [PubMed]
35. Li, B.; Li, H.; Lu, L.; Jiang, J. Structures of human O-GlcNAcase and its complexes reveal a new substrate recognition mode. *Nat. Struct. Mol. Biol.* **2017**, *24*, 362–369. [CrossRef]
36. Elsen, N.L.; Patel, S.B.; Ford, R.E.; Hall, D.L.; Hess, F.; Kandula, H.; Kornienko, M.; Reid, J.; Selnick, H.; Shipman, J.M.; et al. Insights into activity and inhibition from the crystal structure of human O-GlcNAcase. *Nat. Chem. Biol.* **2017**, *13*, 613–615. [CrossRef] [PubMed]
37. Roth, C.; Chan, S.; Offen, W.A.; Hemsworth, G.R.; Willems, L.I.; King, D.T.; Varghese, V.; Britton, R.; Vocadlo, D.J.; Davies, G.J. Structural and functional insight into human O-GlcNAcase. *Nat. Chem. Biol.* **2017**, *13*, 610–612. [CrossRef] [PubMed]
38. Dennis, R.J.; Taylor, E.J.; Macauley, M.S.; Stubbs, K.A.; Turkenburg, J.P.; Hart, S.J.; Black, G.N.; Vocadlo, D.J.; Davies, G.J. Structure and mechanism of a bacterial beta-glucosaminidase having O-GlcNAcase activity. *Nat. Struct. Mol. Biol.* **2006**, *13*, 365–371. [CrossRef]
39. Rao, F.V.; Dorfmueller, H.C.; Villa, F.; Allwood, M.; Eggleston, I.M.; van Aalten, D.M.F. Structural insights into the mechanism and inhibition of eukaryotic O-GlcNAc hydrolysis. *EMBO J.* **2006**, *25*, 1569–1578. [CrossRef] [PubMed]
40. Schimpl, M.; Borodkin, V.S.; Gray, L.J.; van Aalten, D.M. Synergy of peptide and sugar in O-GlcNAcase substrate recognition. *Chem. Biol.* **2012**, *19*, 173–178. [CrossRef] [PubMed]
41. Yang, X.; Su, K.; Roos, M.D.; Chang, Q.; Paterson, A.J.; Kudlow, J.E. O-linkage of N-acetylglucosamine to Sp1 activation domain inhibits its transcriptional capability. *Proc. Natl. Acad. Sci. USA* **2001**, *98*, 6611–6616. [CrossRef]
42. Housley, M.P.; Rodgers, J.T.; Udeshi, N.D.; Kelly, T.J.; Shabanowitz, J.; Hunt, D.F.; Puigserver, P.; Hart, G.W. O-GlcNAc regulates FoxO activation in response to glucose. *J. Biol. Chem.* **2008**, *283*, 16283–16292. [CrossRef]
43. Lewis, B.A.; Hanover, J.A. O-GlcNAc and the epigenetic regulation of gene expression. *J. Biol. Chem.* **2014**, *289*, 34440–34448. [CrossRef]
44. Singh, J.P.; Zhang, K.; Wu, J.; Yang, X. O-GlcNAc signaling in cancer metabolism and epigenetics. *Cancer Lett.* **2015**, *356*, 244–250. [CrossRef] [PubMed]
45. Zhang, K.; Yin, R.; Yang, X. O-GlcNAc: A Bittersweet Switch in Liver. *Front. Endocrinol.* **2014**, *5*, 221. [CrossRef]
46. Taylor, R.P.; Parker, G.J.; Hazel, M.W.; Soesanto, Y.; Fuller, W.; Yazzie, M.J.; McClain, D.A. Glucose Deprivation Stimulates GlcNAc Modification of Proteins through Up-regulation of O-Linked-N-Acetylglucosaminyltransferase. *J. Biol. Chem.* **2008**, *283*, 6050–6057. [CrossRef]
47. Taylor, R.P.; Geisler, T.S.; Chambers, J.H.; McClain, D.A. Up-regulation of O-GlcNAc transferase with glucose deprivation in HepG2 cells is mediated by decreased hexosamine pathway flux. *J. Biol. Chem.* **2009**, *284*, 3425–3432. [CrossRef] [PubMed]
48. Yang, X.; Ongusaha, P.P.; Miles, P.D.; Havstad, J.C.; Zhang, F.; So, W.V.; Kudlow, J.E.; Michell, R.H.; Olefsky, J.M.; Field, S.J.; et al. Phosphoinositide signalling links O-GlcNAc transferase to insulin resistance. *Nature* **2008**, *451*, 964–969. [CrossRef]

49. Kreppel, L.K.; Blomberg, M.A.; Hart, G.W. Dynamic Glycosylation of Nuclear and Cytosolic Proteins: Cloning and characterization of a unique O-GlcNAc transferase with multiple tetratricopeptide repeats. *J. Biol. Chem.* **1997**, *272*, 9308–9315. [CrossRef] [PubMed]
50. Yang, W.H.; Park, S.Y.; Nam, H.W.; Kim, D.H.; Kang, J.G.; Kang, E.S.; Kim, Y.S.; Lee, H.C.; Kim, K.S.; Cho, J.W. NFκB activation is associated with its O-GlcNAcylation state under hyperglycemic conditions. *Proc. Natl. Acad. Sci. USA* **2008**, *105*, 17345–17350. [CrossRef] [PubMed]
51. Dehennaut, V.; Leprince, D.; Lefebvre, T. O-GlcNAcylation, an Epigenetic Mark. Focus on the Histone Code, TET Family Proteins, and Polycomb Group Proteins. *Front. Endocrinol.* **2014**, *5*, 155. [CrossRef] [PubMed]
52. Whelan, S.A.; Dias, W.B.; Thiruneelakantapillai, L.; Lane, M.D.; Hart, G.W. Regulation of insulin receptor substrate 1 (IRS-1)/AKT kinase-mediated insulin signaling by O-Linked beta-N-acetylglucosamine in 3T3-L1 adipocytes. *J. Biol. Chem.* **2010**, *285*, 5204–5211. [CrossRef]
53. Barkovskaya, A.; Prasmickaite, L.; Duveau, D.Y.; Mills, I.G.; Mælandsmo, G.M.; Moestue, S.A.; Itkonen, H.M. Abstract 1131: O-GlcNAc transferase inhibition in breast cancer cells. *Cancer Res.* **2017**, *77*, 1131. [CrossRef]
54. Akella, N.M.; Le Minh, G.; Ciraku, L.; Mukherjee, A.; Bacigalupa, Z.A.; Mukhopadhyay, D.; Sodi, V.L.; Reginato, M.J. O-GlcNAc Transferase Regulates Cancer Stem–like Potential of Breast Cancer Cells. *Mol. Cancer Res.* **2020**, *18*, 585–598. [CrossRef]
55. Ma, Z.; Vosseller, K. Cancer metabolism and elevated O-GlcNAc in oncogenic signaling. *J. Biol. Chem.* **2014**, *289*, 34457–34465. [CrossRef]
56. Chandrashekar, D.S.; Bashel, B.; Balasubramanya, S.A.H.; Creighton, C.J.; Ponce-Rodriguez, I.; Chakravarthi, B.; Varambally, S. UALCAN: A Portal for Facilitating Tumor Subgroup Gene Expression and Survival Analyses. *Neoplasia* **2017**, *19*, 649–658. [CrossRef]
57. Wieland, E.; Rodriguez-Vita, J.; Liebler, S.S.; Mogler, C.; Moll, I.; Herberich, S.E.; Espinet, E.; Herpel, E.; Menuchin, A.; Chang-Claude, J.; et al. Endothelial Notch1 Activity Facilitates Metastasis. *Cancer Cell* **2017**, *31*, 355–367. [CrossRef]
58. Ruland, J. Colon Cancer: Epithelial Notch Signaling Recruits Neutrophils to Drive Metastasis. *Cancer Cell* **2019**, *36*, 213–214. [CrossRef] [PubMed]
59. Lignitto, L.; LeBoeuf, S.E.; Homer, H.; Jiang, S.; Askenazi, M.; Karakousi, T.R.; Pass, H.I.; Bhutkar, A.J.; Tsirigos, A.; Ueberheide, B.; et al. Nrf2 Activation Promotes Lung Cancer Metastasis by Inhibiting the Degradation of Bach1. *Cell* **2019**, *178*, 316–329.e318. [CrossRef]
60. Szymura, S.J.; Zaemes, J.P.; Allison, D.F.; Clift, S.H.; D'Innocenzi, J.M.; Gray, L.G.; McKenna, B.D.; Morris, B.B.; Bekiranov, S.; LeGallo, R.D.; et al. NF-κB upregulates glutamine-fructose-6-phosphate transaminase 2 to promote migration in non-small cell lung cancer. *Cell Commun. Signal.* **2019**, *17*, 24. [CrossRef] [PubMed]
61. Carvalho-cruz, P.; Alisson-Silva, F.; Todeschini, A.R.; Dias, W.B. Cellular glycosylation senses metabolic changes and modulates cell plasticity during epithelial to mesenchymal transition. *Dev. Dyn.* **2018**, *247*, 481–491. [CrossRef] [PubMed]
62. Zhang, X.; Sai, B.; Wang, F.; Wang, L.; Wang, Y.; Zheng, L.; Li, G.; Tang, J.; Xiang, J. Hypoxic BMSC-derived exosomal miRNAs promote metastasis of lung cancer cells via STAT3-induced EMT. *Mol. Cancer* **2019**, *18*, 40. [CrossRef]
63. Diepenbruck, M.; Christofori, G. Epithelial-mesenchymal transition (EMT) and metastasis: Yes, no, maybe? *Curr. Opin. Cell Biol.* **2016**, *43*, 7–13. [CrossRef] [PubMed]
64. Ali, A.; Kim, S.H.; Kim, M.J.; Choi, M.Y.; Kang, S.S.; Cho, G.J.; Kim, Y.S.; Choi, J.-Y.; Choi, W.S. O-GlcNAcylation of NF-κB Promotes Lung Metastasis of Cervical Cancer Cells via Upregulation of CXCR4 Expression. *Mol. Cells* **2017**, *40*, 476–484. [CrossRef]
65. Yan, M.; Xu, Q.; Zhang, P.; Zhou, X.J.; Zhang, Z.Y.; Chen, W.T. Correlation of NF-kappaB signal pathway with tumor metastasis of human head and neck squamous cell carcinoma. *BMC Cancer* **2010**, *10*, 437. [CrossRef]
66. Hu, C.-M.; Tien, S.-C.; Hsieh, P.-K.; Jeng, Y.-M.; Chang, M.-C.; Chang, Y.-T.; Chen, Y.-J.; Chen, Y.-J.; Lee, E.Y.H.P.; Lee, W.-H. High Glucose Triggers Nucleotide Imbalance through O-GlcNAcylation of Key Enzymes and Induces KRAS Mutation in Pancreatic Cells. *Cell Metab.* **2019**, *29*, 1334–1349.e1310. [CrossRef] [PubMed]
67. Kaleem, A.; Ahmad, I.; Hoessli, D.C.; Walker-Nasir, E.; Saleem, M.; Shakoori, A.R.; Nasir ud, D. Epidermal growth factor receptors: Function modulation by phosphorylation and glycosylation interplay. *Mol. Biol. Rep.* **2009**, *36*, 631–639. [CrossRef]
68. Waters, A.M.; Der, C.J. KRAS: The Critical Driver and Therapeutic Target for Pancreatic Cancer. *Cold Spring Harb. Perspect Med.* **2018**, *8*. [CrossRef] [PubMed]
69. da Cunha Santos, G.; Shepherd, F.A.; Tsao, M.S. EGFR mutations and lung cancer. *Annu. Rev. Pathol.* **2011**, *6*, 49–69. [CrossRef]
70. Hsu, W.H.; Yang, J.C.; Mok, T.S.; Loong, H.H. Overview of current systemic management of EGFR-mutant NSCLC. *Ann. Oncol.* **2018**, *29*, i3–i9. [CrossRef] [PubMed]
71. Stateva, S.R.; Villalobo, A. O-GlcNAcylation of the human epidermal growth factor receptor. *Org. Biomol. Chem.* **2015**, *13*, 8196–8204. [CrossRef]
72. Luanpitpong, S.; Angsutararux, P.; Samart, P.; Chanthra, N.; Chanvorachote, P.; Issaragrisil, S. Hyper-O-GlcNAcylation induces cisplatin resistance via regulation of p53 and c-Myc in human lung carcinoma. *Sci. Rep.* **2017**, *7*, 10607. [CrossRef] [PubMed]
73. Wang, D.; Wu, J.; Wang, D.; Huang, X.; Zhang, N.; Shi, Y. Cisplatin enhances protein O-GlcNAcylation by altering the activity of OGT, OGA and AMPK in human non-small cell lung cancer cells. *Int. J. Oncol.* **2021**, *58*, 27. [CrossRef] [PubMed]

74. Rossi, A.; Maio, M.D.; Chiodini, P.; Rudd, R.M.; Okamoto, H.; Skarlos, D.V.; Früh, M.; Qian, W.; Tamura, T.; Samantas, E.; et al. Carboplatin- or Cisplatin-Based Chemotherapy in First-Line Treatment of Small-Cell Lung Cancer: The COCIS Meta-Analysis of Individual Patient Data. *J. Clin. Oncol.* **2012**, *30*, 1692–1698. [CrossRef]
75. Dasari, S.; Tchounwou, P.B. Cisplatin in cancer therapy: Molecular mechanisms of action. *Eur. J. Pharmacol.* **2014**, *740*, 364–378. [CrossRef]
76. Lee, S.J.; Kwon, O.-S. O-GlcNAc Transferase Inhibitor Synergistically Enhances Doxorubicin-Induced Apoptosis in HepG2 Cells. *Cancers* **2020**, *12*, 3154. [CrossRef]
77. Itkonen, H.M.; Minner, S.; Guldvik, I.J.; Sandmann, M.J.; Tsourlakis, M.C.; Berge, V.; Svindland, A.; Schlomm, T.; Mills, I.G. O-GlcNAc transferase integrates metabolic pathways to regulate the stability of c-MYC in human prostate cancer cells. *Cancer Res.* **2013**, *73*, 5277–5287. [CrossRef] [PubMed]
78. Mi, W.; Gu, Y.; Han, C.; Liu, H.; Fan, Q.; Zhang, X.; Cong, Q.; Yu, W. O-GlcNAcylation is a novel regulator of lung and colon cancer malignancy. *Biochim. Biophys. Acta* **2011**, *1812*, 514–519. [CrossRef] [PubMed]
79. Lin, Y.C.; Lin, C.H.; Yeh, Y.C.; Ho, H.L.; Wu, Y.C.; Chen, M.Y.; Chou, T.Y. High O-linked N-acetylglucosamine transferase expression predicts poor survival in patients with early stage lung adenocarcinoma. *Oncotarget* **2018**, *9*, 31032–31044. [CrossRef] [PubMed]
80. Chiaradonna, F.; Ricciardiello, F.; Palorini, R. The Nutrient-Sensing Hexosamine Biosynthetic Pathway as the Hub of Cancer Metabolic Rewiring. *Cells* **2018**, *7*, 53. [CrossRef]
81. Qiang, A.; Slawson, C.; Fields, P.E. The Role of O-GlcNAcylation in Immune Cell Activation. *Front. Endocrinol.* **2021**, *12*. [CrossRef] [PubMed]
82. McClain, D.A.; Lubas, W.A.; Cooksey, R.C.; Hazel, M.; Parker, G.J.; Love, D.C.; Hanover, J.A. Altered glycan-dependent signaling induces insulin resistance and hyperleptinemia. *Proc. Natl. Acad. Sci. USA* **2002**, *99*, 10695–10699. [CrossRef] [PubMed]
83. Machacek, M.; Saunders, H.; Zhang, Z.; Tan, E.P.; Li, J.; Li, T.; Villar, M.T.; Artigues, A.; Lydic, T.; Cork, G.; et al. Elevated O-GlcNAcylation enhances pro-inflammatory Th17 function by altering the intracellular lipid microenvironment. *J. Biol. Chem.* **2019**, *294*, 8973–8990. [CrossRef] [PubMed]
84. Liu, B.; Salgado, O.C.; Singh, S.; Hippen, K.L.; Maynard, J.C.; Burlingame, A.L.; Ball, L.E.; Blazar, B.R.; Farrar, M.A.; Hogquist, K.A.; et al. The lineage stability and suppressive program of regulatory T cells require protein O-GlcNAcylation. *Nat. Commun.* **2019**, *10*, 354. [CrossRef] [PubMed]
85. Chang, Y.H.; Weng, C.L.; Lin, K.I. O-GlcNAcylation and its role in the immune system. *J. Biomed. Sci.* **2020**, *27*, 57. [CrossRef]
86. Golks, A.; Tran, T.T.; Goetschy, J.F.; Guerini, D. Requirement for O-linked N-acetylglucosaminyltransferase in lymphocytes activation. *EMBO J.* **2007**, *26*, 4368–4379. [CrossRef] [PubMed]
87. Swamy, M.; Pathak, S.; Grzes, K.M.; Damerow, S.; Sinclair, L.V.; van Aalten, D.M.; Cantrell, D.A. Glucose and glutamine fuel protein O-GlcNAcylation to control T cell self-renewal and malignancy. *Nat. Immunol.* **2016**, *17*, 712–720. [CrossRef]
88. Hwang, J.S.; Kim, K.H.; Park, J.; Kim, S.M.; Cho, H.; Lee, Y.; Han, I.O. Glucosamine improves survival in a mouse model of sepsis and attenuates sepsis-induced lung injury and inflammation. *J. Biol Chem* **2019**, *294*, 608–622. [CrossRef] [PubMed]
89. Shapouri-Moghaddam, A.; Mohammadian, S.; Vazini, H.; Taghadosi, M.; Esmaeili, S.A.; Mardani, F.; Seifi, B.; Mohammadi, A.; Afshari, J.T.; Sahebkar, A. Macrophage plasticity, polarization, and function in health and disease. *J. Cell Physiol.* **2018**, *233*, 6425–6440. [CrossRef]
90. He, Y.; Ma, X.; Li, D.; Hao, J. Thiamet G mediates neuroprotection in experimental stroke by modulating microglia/macrophage polarization and inhibiting NF-κB p65 signaling. *J. Cereb. Blood Flow Metab.* **2017**, *37*, 2938–2951. [CrossRef]
91. Li, T.; Li, X.; Attri, K.S.; Liu, C.; Li, L.; Herring, L.E.; Asara, J.M.; Lei, Y.L.; Singh, P.K.; Gao, C.; et al. O-GlcNAc Transferase Links Glucose Metabolism to MAVS-Mediated Antiviral Innate Immunity. *Cell Host Microbe* **2018**, *24*, 791–803.e796. [CrossRef] [PubMed]
92. Song, N.; Qi, Q.; Cao, R.; Qin, B.; Wang, B.; Wang, Y.; Zhao, L.; Li, W.; Du, X.; Liu, F.; et al. MAVS O-GlcNAcylation Is Essential for Host Antiviral Immunity against Lethal RNA Viruses. *Cell Rep.* **2019**, *28*, 2386–2396.e2385. [CrossRef] [PubMed]
93. Allison, D.F.; Wamsley, J.J.; Kumar, M.; Li, D.; Gray, L.G.; Hart, G.W.; Jones, D.R.; Mayo, M.W. Modification of RelA by O-linked N-acetylglucosamine links glucose metabolism to NF-κB acetylation and transcription. *Proc. Natl. Acad. Sci. USA* **2012**, *109*, 16888–16893. [CrossRef]
94. Li, X.; Gong, W.; Wang, H.; Li, T.; Attri, K.S.; Lewis, R.E.; Kalil, A.C.; Bhinderwala, F.; Powers, R.; Yin, G.; et al. O-GlcNAc Transferase Suppresses Inflammation and Necroptosis by Targeting Receptor-Interacting Serine/Threonine-Protein Kinase 3. *Immunity* **2019**, *50*, 576–590.e576. [CrossRef]
95. Hou, C.-W.; Mohanan, V.; Zachara, N.E.; Grimes, C.L. Identification and biological consequences of the O-GlcNAc modification of the human innate immune receptor, Nod2. *Glycobiology* **2015**, *26*, 13–18. [CrossRef] [PubMed]
96. Andrejeva, G.; Rathmell, J.C. Similarities and Distinctions of Cancer and Immune Metabolism in Inflammation and Tumors. *Cell Metab.* **2017**, *26*, 49–70. [CrossRef] [PubMed]
97. Thorsson, V.; Gibbs, D.L.; Brown, S.D.; Wolf, D.; Bortone, D.S.; Ou Yang, T.H.; Porta-Pardo, E.; Gao, G.F.; Plaisier, C.L.; Eddy, J.A.; et al. The Immune Landscape of Cancer. *Immunity* **2018**, *48*, 812–830.e814. [CrossRef]
98. Chanmee, T.; Ontong, P.; Konno, K.; Itano, N. Tumor-Associated Macrophages as Major Players in the Tumor Microenvironment. *Cancers* **2014**, *6*, 1670. [CrossRef]

99. Zheng, X.; Mansouri, S.; Krager, A.; Grimminger, F.; Seeger, W.; Pullamsetti, S.S.; Wheelock, C.E.; Savai, R. Metabolism in tumour-associated macrophages: A quid pro quo with the tumour microenvironment. *Eur. Respir. Rev.* **2020**, *29*. [CrossRef] [PubMed]
100. Liu, J.; Geng, X.; Hou, J.; Wu, G. New insights into M1/M2 macrophages: Key modulators in cancer progression. *Cancer Cell Int.* **2021**, *21*, 389. [CrossRef]
101. Cassetta, L.; Pollard, J.W. Targeting macrophages: Therapeutic approaches in cancer. *Nat. Rev. Drug Discov.* **2018**, *17*, 887–904. [CrossRef] [PubMed]
102. Prenen, H.; Mazzone, M. Tumor-associated macrophages: A short compendium. *Cell Mol. Life Sci.* **2019**, *76*, 1447–1458. [CrossRef]
103. Waldman, A.D.; Fritz, J.M.; Lenardo, M.J. A guide to cancer immunotherapy: From T cell basic science to clinical practice. *Nat. Rev. Immunol.* **2020**, *20*, 651–668. [CrossRef] [PubMed]
104. Ramakrishnan, P.; Clark, P.M.; Mason, D.E.; Peters, E.C.; Hsieh-Wilson, L.C.; Baltimore, D. Activation of the transcriptional function of the NF-κB protein c-Rel by O-GlcNAc glycosylation. *Sci. Signal.* **2013**, *6*, ra75. [CrossRef] [PubMed]
105. Jiang, M.; Xu, B.; Li, X.; Shang, Y.; Chu, Y.; Wang, W.; Chen, D.; Wu, N.; Hu, S.; Zhang, S.; et al. O-GlcNAcylation promotes colorectal cancer metastasis via the miR-101-O-GlcNAc/EZH2 regulatory feedback circuit. *Oncogene* **2019**, *38*, 301–316. [CrossRef]
106. Itkonen, H.M.; Poulose, N.; Steele, R.E.; Martin, S.E.S.; Levine, Z.G.; Duveau, D.Y.; Carelli, R.; Singh, R.; Urbanucci, A.; Loda, M.; et al. Inhibition of O-GlcNAc transferase renders prostate cancer cells dependent on CDK9. *Mol. Cancer Res.* **2020**. [CrossRef] [PubMed]
107. Liu, Y.; Huang, H.; Cao, Y.; Wu, Q.; Li, W.; Zhang, J. Suppression of OGT by microRNA24 reduces FOXA1 stability and prevents breast cancer cells invasion. *Biochem. Biophys. Res. Commun.* **2017**, *487*, 755–762. [CrossRef] [PubMed]
108. Itkonen, H.M.; Gorad, S.S.; Duveau, D.Y.; Martin, S.E.; Barkovskaya, A.; Bathen, T.F.; Moestue, S.A.; Mills, I.G. Inhibition of O-GlcNAc transferase activity reprograms prostate cancer cell metabolism. *Oncotarget* **2016**, *7*, 12464–12476. [CrossRef]
109. Xu, W.; Zhang, X.; Wu, J.L.; Fu, L.; Liu, K.; Liu, D.; Chen, G.G.; Lai, P.B.; Wong, N.; Yu, J. O-GlcNAc transferase promotes fatty liver-associated liver cancer through inducing palmitic acid and activating endoplasmic reticulum stress. *J. Hepatol.* **2017**, *67*, 310–320. [CrossRef] [PubMed]
110. Trapannone, R.; Rafie, K.; van Aalten, D.M. O-GlcNAc transferase inhibitors: Current tools and future challenges. *Biochem. Soc. Trans.* **2016**, *44*, 88–93. [CrossRef] [PubMed]
111. Borodkin, V.S.; Schimpl, M.; Gundogdu, M.; Rafie, K.; Dorfmueller, H.C.; Robinson, D.A.; van Aalten, D.M.F. Bisubstrate UDP-peptide conjugates as human O-GlcNAc transferase inhibitors. *Biochem. J.* **2014**, *457*, 497–502. [CrossRef] [PubMed]

cancers

MDPI

Article

Comparative *O*-GlcNAc Proteomic Analysis Reveals a Role of *O*-GlcNAcylated SAM68 in Lung Cancer Aggressiveness

Chia-Hung Lin [1,†], Chen-Chung Liao [2,†], Shu-Ying Wang [3], Chia-Yi Peng [4], Yi-Chen Yeh [1], Mei-Yu Chen [4,5,6,*] and Teh-Ying Chou [1,3,4,6,*]

1 Department of Pathology and Laboratory Medicine, Taipei Veterans General Hospital, Taipei 11227, Taiwan; chiahunglin0222@gmail.com (C.-H.L.); ycyeh2@vghtpe.gov.tw (Y.-C.Y.)
2 Metabolomics-Proteomics Research Center, National Yang Ming Chiao Tung University, Taipei 11221, Taiwan; ccliao@nycu.edu.tw
3 Institute of Clinical Medicine, College of Medicine, National Yang Ming Chiao Tung University, Taipei 11221, Taiwan; sywang15@vghtpe.gov.tw
4 Institute of Biochemistry and Molecular Biology, College of Life Sciences, National Yang Ming Chiao Tung University, Taipei 11221, Taiwan; ivy110024@gmail.com
5 Faculty of Medicine, School of Medicine, National Yang Ming Chiao Tung University, Taipei 11221, Taiwan
6 Cancer Progression Research Center, National Yang Ming Chiao Tung University, Taipei 11221, Taiwan
* Correspondence: meychen@nycu.edu.tw (M.-Y.C.); tychou@vghtpe.gov.tw (T.-Y.C.)
† These authors contributed equally to this work.

Simple Summary: Lung cancer claims the most lives annually among cancers; to date, invasion and metastasis still pose challenges to effective treatment. *O*-GlcNAcylation, an enzymatic modification of proteins after biosynthesis, modulates the functions of many proteins. Aberrant *O*-GlcNAcylation is linked to pathogenic mechanisms of cancer, including invasion and metastasis. However, little is known about the profile of *O*-GlcNAcylated proteins involved in cancer aggressiveness. Here, by comparing profiles of *O*-GlcNAcylated proteins from two lung cancer cell lines different in their invasive potential, we identified candidates for *O*-GlcNAcylated proteins that may be involved in cancer aggressiveness. One of these candidates, SAM68, was further characterized. Results confirmed *O*-GlcNAcylation of SAM68; functional analyses on SAM68 with mutations at *O*-GlcNAcylation sites suggested a role of *O*-GlcNAcylated SAM68 in modulating lung cancer cell migration/invasion. Future elucidation of the functional significance of differential *O*-GlcNAcylation of proteins identified in this study may provide new insights into mechanisms of lung cancer progression.

Abstract: *O*-GlcNAcylation is a reversible and dynamic post-translational protein modification catalyzed by *O*-GlcNAc transferase (OGT). Despite the reported association of *O*-GlcNAcylation with cancer metastasis, the *O*-GlcNAc proteome profile for cancer aggressiveness remains largely uncharacterized. Here, we report our comparative *O*-GlcNAc proteome profiling of two differentially invasive lung adenocarcinoma cell lines, which identified 158 down-regulated and 106 up-regulated candidates in highly invasive cells. Among these differential proteins, a nuclear RNA-binding protein, SAM68 (SRC associated in mitosis of 68 kDa), was further investigated. Results showed that SAM68 is *O*-GlcNAcylated and may interact with OGT in the nucleus. Eleven *O*-GlcNAcylation sites were identified, and data from mutant analysis suggested that multiple serine residues in the *N*-terminal region are important for *O*-GlcNAcylation and the function of SAM68 in modulating cancer cell migration and invasion. Analysis of clinical specimens found that high SAM68 expression was associated with late cancer stages, and patients with high-OGT/high-SAM68 expression in their tumors had poorer overall survival compared to those with low-OGT/low-SAM68 expression. Our study revealed an invasiveness-associated *O*-GlcNAc proteome profile and connected *O*-GlcNAcylated SAM68 to lung cancer aggressiveness.

Keywords: invasiveness; lung adenocarcinoma; *O*-GlcNAc transferase; *O*-GlcNAcylation; SAM68

Citation: Lin, C.-H.; Liao, C.-C.; Wang, S.-Y.; Peng, C.-Y.; Yeh, Y.-C.; Chen, M.-Y.; Chou, T.-Y. Comparative *O*-GlcNAc Proteomic Analysis Reveals a Role of *O*-GlcNAcylated SAM68 in Lung Cancer Aggressiveness. *Cancers* **2022**, *14*, 243. https://doi.org/10.3390/cancers14010243

Academic Editors: Ikram El Yazidi-Belkoura and Tony Lefebvre

Received: 11 November 2021
Accepted: 30 December 2021
Published: 4 January 2022

1. Introduction

Lung cancer is by far the most common cause of cancer death worldwide; globally reported deaths from lung cancer were around 180,000 in 2020 [1]. The two major diagnosed classes of lung cancers are non-small-cell lung cancer (NSCLC, ~85%) and small-cell lung cancer (~15%) [2]. Adenocarcinoma accounting for ~40% and ~70% of the male and female lung cancers in Taiwan, is the prevailing histological type of NSCLC and the most common primary lung malignancy in patients who have never smoked [2–4]. The prognosis of lung cancer remains poor, and the overall 5-year survival rate is only 10–20% [1]. The mortality mainly results from metastasis, which is a progression requiring cancer cells to undergo intravasation, survival in circulation, extravasation, and colonization [5]. Cancer invasion—a process involving disruption of the surrounding extracellular matrix and increasing cell motility—is the initial and a critical step for metastasis. However, a complete understanding of the molecular mechanisms underlying the regulation of cancer invasiveness remains elusive.

O-GlcNAcylation is the addition of a single β-*N*-acetylglucosamine (GlcNAc) to the hydroxyl group of serine/threonine residues in many nuclear and cytoplasmic proteins [6]. Cycling of this *O*-linked GlcNAc (*O*-GlcNAc) modification is regulated by a pair of evolutionary conserved enzymes: *O*-GlcNAc transferase (OGT) catalyzes the attachment of the GlcNAc moiety from UDP-GlcNAc to target proteins [7,8], whereas *O*-GlcNAcase (OGA) removes *O*-GlcNAc from *O*-GlcNAcylated proteins [9,10]. *O*-GlcNAcylation is an abundant and dynamic post-translational modification involved in many crucial cellular processes, including regulation of gene expression, signal transduction, metabolism, and cell cycles [11–14]. Ever-growing evidence indicates that dysregulation of cellular *O*-GlcNAc levels is associated with various diseases, including diabetes, neurodegenerative disorders, and cancer [13–16].

Aberrant *O*-GlcNAcylation has been linked to pathogenic mechanisms of cancer, including cell proliferation, survival, invasion, and metastasis [17]. Regulation of *O*-GlcNAc modification on specific proteins can modulate the migration and invasiveness of tumor cells. For instance, *O*-GlcNAcylation on Ser-112 of Snail, an important transcription factor involved in promoting the epithelial-mesenchymal transition and cancer cell invasiveness, can increase the stability of Snail protein and promote tumor invasion through repression of E-cadherin expression [18]. In breast cancer cells, loss of *O*-GlcNAcylation on Ser-108 of cofilin results in destabilization of invadopodia and impairs cancer cell invasion [19], and *O*-GlcNAcylation on Ser-408 of TAK1 binding protein 3 (TAB3) contributes to TAB3-mediated promotion of cancer cell migration and invasion by activating NF-kB [20]. Nevertheless, an overall profile of *O*-GlcNAcylated proteins that are involved in the pathogenesis of cancer, especially in invasion/metastasis, is still unclear.

We previously found that OGT expression is an independent prognostic factor in patients with lung adenocarcinoma [21]. It has been reported that knockdown of OGT expression in A549 lung cancer cells decreases their invasiveness in vitro [22]. TGF-β-induced epithelial-mesenchymal transition of A549 cells is accompanied by significantly increased OGT expression and cellular *O*-GlcNAcylation [23]. These observations suggest that *O*-GlcNAcylated proteins likely contribute to lung cancer invasion/metastasis. In light of this, this study aimed at uncovering *O*-GlcNAcylated proteins that modulate the invasiveness of lung cancer cells. We employed two lung adenocarcinoma cell lines with different invasive abilities; using label-free Mass spectrometry (MS) to analyze wheat germ agglutinin (WGA)-enriched proteins from these cells, we profiled the *O*-GlcNAc proteomes to identify *O*-GlcNAcylated proteins with differential abundance in the two cell lines. Among the identified proteins, we further investigated the SRC associated in mitosis of the 68-kDa (SAM68) protein to characterize its *O*-GlcNAcylation and its role in the migration/invasion of lung cancer cells.

2. Materials and Methods

2.1. Cell Lines and Cell Culture

Human lung adenocarcinoma cell lines CL1-1 and CL1-5 and A549 cells were maintained in RPMI-1640 (Gibco, Grand Island, NY, USA) supplemented with heat-inactivated 10% (*v*/*v*) fetal bovine serum (FBS) and 1% (*v*/*v*) penicillin/streptomycin. Human embryonic kidney 293T cells were maintained in Dulbecco's modified Eagle's medium (DMEM; Gibco, Grand Island, NY, USA) containing 10% (*v*/*v*) FBS. CL1-1 and CL1-5 cells were kindly provided by Prof. Pan-Chyr Yang from National Taiwan University, Taiwan. A549 and 293T cells, which were originally from Bioresource Collection and Research Center, Taiwan, were kindly gifted by Prof. Fung-Fang Wang from National Yang-Ming University, Taiwan. All cells were cultured at 37 °C in a humidified atmosphere with 5% CO_2.

2.2. Nuclear/Cytosolic Fractionation

CL1-1 and CL1-5 cells were trypsinized to allow detachment from culture dishes, washed twice with PBS, and lysed in solution A (20 mM Tris, pH 7.5, 3 mM $MgCl_2$, 1 mM EDTA, 0.1% Triton X-100, and protease inhibitors) at 1×10^7 cells/mL on ice for 30 min. Lysates were centrifuged at 1500× *g* for 5 min, and supernatants were collected as the cytosolic fractions. Pellets were washed twice with solution A, suspended, vortexed in solution B (20 mM HEPES, pH 7.9, 25% Glycerol, 1% Triton X-100, and protease inhibitors), and incubated on ice for 30 min; after centrifugation at 15,000× *g* for 15 min, supernatants were collected as the nuclear fractions.

2.3. Enrichment of O-GlcNAcylated Proteins

For the enrichment of *O*-GlcNAcylated proteins, we used the Glycoprotein Isolation Kit (WGA #89805, Thermo Scientific, Waltham, MA, USA) following the manufacturer's protocol. Briefly, 50% wheat germ agglutinin (WGA) resin slurry (200 μL/column) was applied to spin columns and washed twice with the Binding/Wash Buffer. Samples of nuclear or cytosolic fractions (1.5 mg/column) were added to the columns and incubated for 15 min at room temperature (RT) with end-over-end mixing using a rotator. WGA resin was collected by centrifugation and washed four times with the Binding/Wash Buffer. WGA-captured proteins were eluted twice by incubating the resin with 200 μL of elution buffer for 10 min and spinning at 1000× *g* for 1 min, and eluate from the same column was pooled and concentrated by using an Amicon Ultra-0.5, 3 KDa (Millipore, Billerica, MA, USA). WGA-captured glycoproteins were quantified using Bradford assays (Bio-Rad Laboratories, Richmond, CA, USA).

2.4. In-Gel Digestion and Liquid Chromatography (LC)-Mass Spectrometry (MS)/MS Analysis

Proteins eluted from WGA resin were separated by 10% SDS-PAGE and stained using the VisPRO 5 min Protein Stain Kit (Visual Protein Biotechnology Corporation, Taipei, Taiwan). Each sample lane was excised, cut into pieces, and dried in a SpeedVac (Thermo Electron, Waltham, MA, USA). Dried gel pieces of each sample were first incubated in 1% β-mercaptoethanol in 25 mM NH_4HCO_3 for 20 min at RT in the dark and further incubated for 20 min after an equal volume of 5% 4-vinylpyridine in 25 mM NH_4HCO_3 and 50% acetonitrile was added. After the removal of solvent, gel slices were washed in 25 mM NH_4HCO_3 for 10 min and dried in a SpeedVac. For in-gel digestion, gel pieces were incubated with cold 0.1% modified trypsin (Promega, Madison, WI, USA) first at 4 °C for 20 min and subsequently at 37 °C overnight. The digest solution containing tryptic peptides was collected, dried in a Speed-Vac, and stored at −20 °C until further analysis.

Each trypsin-digested sample was suspended in 20 μL of 0.1% formic acid and analyzed using the nanoACQUITY™ system (Waters, Milford, MA, USA), which was connected to an Orbitrap Elite hybrid mass spectrometer equipped with a nanoelectrospray ionization source (Thermo Scientific, Waltham, MA, USA). Peptides were separated by reverse-phase LC using a BEH-C18 column (25 cm × 75 μm, Waters, Milford, MA, USA) with a segmented gradient of 5% to 35% acetonitrile in 0.1% formic acid for 210 min at a flow rate of 300 nl/min. Eluted peptides were ionized at a spray voltage of 1.7 kV and introduced into the Orbitrap Elite for MS analysis in the positive ion mode with data-dependent acquisition (2.0 Da isolation width). A full MS scan was set at a resolution of 30,000 at m/z 400. Spectrum data of peptides were obtained by full-mass survey scan (m/z range: 350–1600). MS/MS scan by collision-induced dissociation was performed on 10 most intense multiply charged ions (2^+ and 3^+).

The mass spectrometry proteomics data have been deposited in the ProteomeXchange Consortium (http://proteomecentral.proteomexchange.org, accessed on 10 November 2021) via the PRIDE [24] partner repository with the dataset identifier PXD029627 and DOI 10.6019/PXD029627.

2.5. Protein Identification and Analysis

For protein identification, the acquired MS data were analyzed by using the Peaks7.5 Studio software for proteomics (Bioinformatics Solutions, Waterloo, Canada) to search against the UniProt human protein database (containing 192,901 sequences; released in January 2021; http://www.uniprot.org/, accessed on 15 January 2021). Search parameters were as follows: enzyme, trypsin; parent and fragment mass error tolerance, 50 ppm and 0.8 Da, respectively; allowing two missed cleavages, oxidation on methionine (+15.99 Da), and carbamidomethylation on cysteine (+57.02 Da) as variable modifications. The average local confidence (ALC) was >80%. A decoy database was constructed to estimate the false discovery rate (FDR), and FDR was controlled at <0.1%. A protein was identified when at least one unique peptide was matched. Protein quantitation based on the MS spectra was performed with in-house software [25]. *O*-GlcNAcylated peptide sequences and sites of WGA-captured proteins were identified using the PeaksPTM module of the PEAKS 7.5 software (Bioinformatics Solutions, Waterloo, Canada). Possible ions (e.g., b, y, y-NH_3, or b-H_2O) of modified fragment peptides in the MS spectra were manually labeled. Differential WGA-bound nuclear and cytosolic glycoproteins were analyzed using the PANTHER classification system (http://www.pantherdb.org/, accessed on 13 February 2019) [26] for molecular function, biological process, and protein class-based gene ontology. Pathway analysis was performed using Ingenuity Pathway Analysis (IPA; Ingenuity Systems, Redwood City, CA, USA) based on experimental observations of the target genes reported in the literature to determine differentially regulated signaling networks with significance calculated using Fisher's exact test (p-value < 0.05).

2.6. Western Analysis

Samples of nuclear and cytosolic lysates and WGA flow-through, wash, and elution fractions were resolved in 10% SDS-PAGE gels and transferred onto polyvinylidene fluoride transfer membranes (PVDF, Pall Corporation, East Hills, NY, USA). Membranes were blocked in TBS-T buffer containing 5% BSA for 1 h at RT and were then incubated with primary antibodies overnight at 4 °C. Antibodies used include: anti-*O*-GlcNAc CTD110.6 (1:2000; MMS-248R; BioLegend, CA, USA), anti-IQGAP1 (1:1000; sc-10792; Santa Cruz Biotechnology Inc., Santa Cruz, CA, USA), anti-LSD1 (1:1000; C69G12; Cell Signaling Technology, Beverly, MA, USA), anti-MTA2 (1:1000; ab50209; Abcam, Cambridge, MA, USA), anti-SAM68 (1:5000; sc-333; Santa Cruz Biotechnology, Inc.), anti-B23 (1:1000; sc-6013-R; Santa Cruz Biotechnology Inc.), anti-Rock1 (1:1000; sc-17794; Santa Cruz Biotechnology Inc.), anti-OGT (1:2000; 11576-2-AP; Proteintech, Rosemont, IL, USA), anti-PARP1 (1:1000; GTX100573; GeneTex, Hsinchu City, Taiwan), anti-β-Actin (1:20000; NB600-501; Novus Biologicals, Centennial, CO. USA), anti-α-Tubulin (1:1000; T9026; Sigma-Aldrich, St. Louis, MO, USA), and anti-Flag (1:1000; F3165; Sigma-Aldrich). Rabbit anti-α-Enolase antibody was kindly provided by Prof. Neng-Yao Shih from the National Health Research Institutes, Taiwan. After washing, membranes were incubated with appropriate horseradish peroxidase HRP-conjugated secondary antibodies for 1 h at RT, and protein signals were visualized using a chemiluminescence ECL kit (HyCell International Co., Ltd., Taipei, Taiwan).

2.7. Co-Immunoprecipitation

CL1-5 cells were washed in PBS twice and treated with or without 1 mM crosslinker dithiobis (succinimidyl propionate) (DSP; Thermo Scientific #22585) for 15 min at RT. The reaction was quenched in 20 mM Tris-base pH 7.5 for 10 min at RT. Cells were harvested and suspended in the lysis buffer (150 mM NaCl, 0.5% Sodium deoxycholate, 1% NP-40, 0.1% SDS) with a protease inhibitor cocktail, gently passed through a 26-gauge needle 20 times, vortexed for 90 s, and then incubated on ice for 30 min. After centrifugation at 12,500 rpm in a microfuge for 10 min at 4 °C, the supernatant was collected and the protein concentration was measured by Bradford assays (Bio-Rad Laboratories, Richmond, CA, USA). Then, 20 μL of protein G magnetic beads was incubated with 1 mg of lysates and 1 μg of anti-OGT, anti-SAM68 or control IgG antibodies at 4 °C overnight. The beads were washed with lysis buffer three times and boiled in β-mercaptoethanol-containing sample buffer at 100 °C for 10 min, and proteins were subjected to SDS-PAGE and Western analysis.

2.8. Immunofluorescence Staining

CL1-5 cells were transiently transfected with a plasmid expressing Flag-tagged SAM68 or Myc-tagged OGT. Cells cultured on coverslips were fixed with 4% paraformaldehyde in PBS for 10 min, permeabilized with 0.5% Triton X-100 in PBS for 5 min, and blocked with 10% FBS in PBS for 30 min at RT. After PBS washing, fixed cells were incubated with primary antibodies, including anti-OGT (1:200; 11576-2-AP; ProteinTech), anti-SAM68 (1:200; sc-333; Santa Cruz Biotechnology Inc.), anti-Myc (1:100; MMS-150P; Covance, Richmond, CA, USA) or anti-Flag (20 μg/mL; F3165; Sigma-Aldrich), at 4 °C overnight. Cells were washed again in PBS and incubated with DAPI and appropriate rhodamine- or FITC-conjugated secondary antibodies at RT for 1 h. Finally, the slides were washed in PBS and mounted in mounting solution. Cells were examined under a confocal fluorescence microscope.

2.9. Knockdown of SAM68-Encoding KHDRBS1 by Lentivirus-Delivered shRNAs

Plasmids for the expression of shRNAs targeting *KHDRBS1* (which encodes SAM68) (TRCN0000000044, designated as #1, and TRCN0000000048, designated as #2) were obtained from the National RNAi Core Facility Platform located at the Institute of Molecular Biology/Genomic Research Center, Academia Sinica, Taipei, Taiwan. Virus packaging and target cell transduction were performed as described previously [27].

2.10. Transwell Migration and Invasion Assays

Cell culture inserts with 8-μm pores (Millicell®; Merck Millipore Ltd., Billerica, MA, USA) were used. For migration assay, 2×10^4 cells/well in 1% FBS-containing medium were placed in the upper chambers; the lower chambers were filled with 10% FBS-containing medium, and the assay was performed at 37 °C for 6 h. For invasion assay, the cell culture inserts were coated with matrigel (40 μg/well), and 2×10^4 cells/well in 10% FBS-containing medium were seeded on top of each filter insert; medium supplemented with 10% serum was added to the lower chambers, and the assay was performed at 37 °C for 24 h. At the end of the assays, the inserts were fixed in 100% methanol for 20 min, air dried, and stained with Giemsa Stain (Sigma-Aldrich) overnight. The inner side of the inserts was wiped with cotton swaps, and the cells that migrated or invaded through the insert were counted under a light microscope.

2.11. Immunohistochemical (IHC) Staining

A tissue microarray (TMA) that contained 174 patients with lung adenocarcinoma at various stages was analyzed; these patients underwent tumor resection at Taipei Veterans General Hospital (Taipei-VGH) between 2002 and 2006. The collection and usage of clinical samples complied with the regulations of the Taipei-VGH Institutional Review Board (IRB No. 2021-04-011BCF). The stage of lung adenocarcinoma was determined according to the Union for International Cancer Control/American Joint Committee on Cancer TNM classification. IHC staining was performed as described previously [21]. Briefly, after rehydration, antigen retrieval, and peroxidase blocking, TMA sections were incubated with anti-SAM68 (1:50; LifeSpan BioSciences, Inc., Seattle, WA, USA), anti-OGT (1:50; ProteinTech, Chicago, IL, USA) or anti-*O*-GlcNAc (1:200; Thermo Fisher Scientific, Waltham, MA, USA) antibodies overnight at 4 °C and subsequently, after washing in phosphate-buffer saline (PBS), with peroxidase-labeled secondary antibody for 1 h at RT. Sections were then incubated with diaminobenzidine, washed, and counterstained with hematoxylin before being mounted. The staining was examined by pathologists and semi-quantitatively scored as follows: 0 (no staining), 1 (weakly positive), 2 (moderately positive), and 3 (strongly positive); percentage scores were 0–100%. The IHC total score for each specimen was the intensity score multiplied by the percentage score. The median of the total scores was used as the cut-off value to categorize patients into high- and low-expression groups.

2.12. Statistical Analysis

Presented quantitative results are the mean ± standard deviation from at least three independent experiments. Comparisons were performed by two-tailed unpaired Student's *t*-test. A p value < 0.05 was considered statistically significant.

3. Results

3.1. Differential O-GlcNAcylated Proteins in Lung Adenocarcinoma Cell Lines with Low and High Invasiveness

Aiming at identifying *O*-GlcNAcylated proteins related to cancer invasiveness, we employed CL1-1 and CL1-5, which are two well-established human lung adenocarcinoma cell lines with differential invasiveness (Figure 1A). These cell lines have been established from one human lung adenocarcinoma clone through sequential in vitro Transwell enrichment for invasive subpopulations; CL1-1 was established after one round while CL1-5 after five rounds of selection [28]. CL1-1 cells exhibit a typical epithelial-type morphology and tend to form cell clusters, while CL1-5 cells display a spindle-shaped fibroblast-like morphology with less cohesiveness; accordingly, CL1-1 cells have low invasiveness while CL1-5 cells are highly invasive. The isogenic nature of these two cell lines offers an advantage in this comparative study for invasiveness-associated proteomes, as proteomic differences caused by distinct genetic backgrounds would be minimized. When cytosolic and nuclear fractions of lysates from CL1-1 and CL1-5 cells were examined for *O*-GlcNAcylated proteins by Western analysis using *O*-GlcNAc-recognizing antibodies, some distinct differential *O*-GlcNAcylated protein bands were noted (Figure 1B), indicating that cells with different degrees of invasiveness may have differential *O*-GlcNAc proteomes.

3.2. Identification and Validation of Differential WGA-Bound Glycoproteins in Lung Adenocarcinoma Cell Lines with Low and High Invasiveness

Towards identifying *O*-GlcNAcylated proteins with differential amounts in two cell lines, we employed WGA to perform enrichment of *O*-GlcNAcylated proteins. Western analysis showed that *O*-GlcNAcylated proteins from nuclear/cytosolic fractions were indeed enriched in the WGA elution fractions (Figures S1A,B and S14). We next performed in-gel trypsin digestion on samples of WGA elution fractions, and the resulting tryptic peptide mixtures from CL1-1 and CL1-5 samples were independently analyzed in triplicate by nano-LC-MS/MS for protein identification. Comparing the mass spectrometry data and using a threshold of >1.5-fold change between the two cell lines, we reproducibly identified 63 down-regulated and 62 up-regulated candidates for nuclear *O*-GlcNAcylated proteins, and 95 down-regulated and 44 up-regulated cytosolic ones in CL1-5 cells (Student's *t*-test, p-value < 0.05) (Figure 2A and Tables S1 and S2).

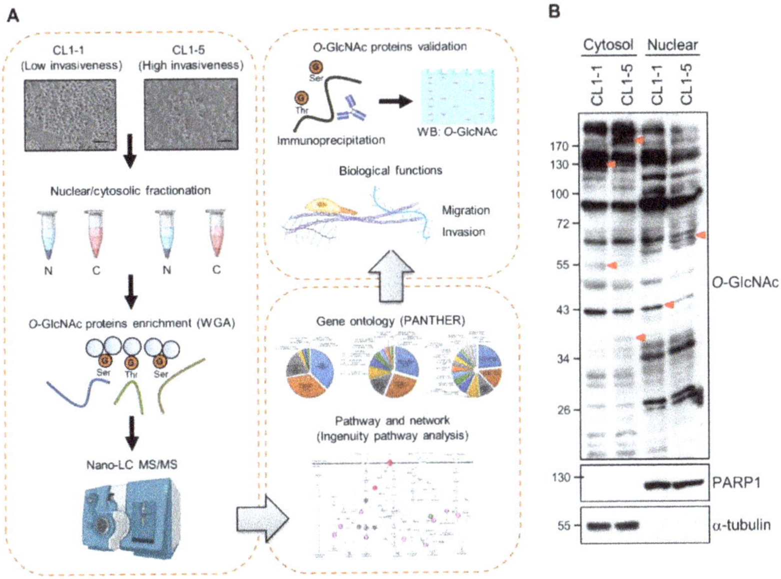

Figure 1. Profiling of the *O*-GlcNAc proteomes of two differentially invasive lung adenocarcinoma cell lines. (**A**) A schematic workflow of this study. Lysates were prepared from subconfluent lowly invasive CL1-1 and highly invasive CL1-5 cells and fractionated into nuclear (N) and cytosolic (**C**) extracts. *O*-GlcNAcylated proteins were enriched using resin-bound wheat germ agglutinin (WGA). Tryptic peptides of WGA-bound proteins were subjected to nano-LC-MS/MS. Reproducibly identified differential WGA-bound proteins were further analyzed in silico using bioinformatics tools including PANTHER Gene Ontology and Ingenuity pathway analysis. Selected differential WGA-bound proteins were further validated and characterized for roles in cancer cell migration and invasion. (**B**) Representative results of Western analysis of cytosolic and nuclear exacts of CL1-1 and CL1-5 cells. Proteins were separated by 10% SDS-PAGE and subjected to Western blotting using antibodies for *O*-GlcNAc, PARP1 (a nuclear protein marker) and α-tubulin (a cytosolic protein marker). Arrowheads point to examples of differential bands between CL1-1 and CL1-5 samples. Detailed information about the Western blotting can be found in Figure S7. PANTHER, Protein ANalysis THrough Evolutionary Relationships; MS, Mass Spectrometry; orange circled G, *O*-GlcNAc.

We further performed gene ontology (GO) analysis and Ingenuity Pathway Analysis (IPA) to identify functional networks and canonical pathways of these differential WGA-bound nuclear and cytosolic proteins. Results of GO slim analysis showed that the top identified protein classes were nucleic acid binding proteins, followed by cytoskeletal proteins, hydrolases, and enzyme modulators (Figure S2A); the top associated molecular functions were binding and catalytic activities (Figure S2B), and these proteins were mainly involved in the biological functions of metabolism, cellular process, and cellular component organization or biogenesis (Figure S2C). Results of IPA revealed canonical pathways and network functions that were significantly associated with the differential WGA-bound proteins (Figure 2B and Table S3). Interestingly, the top associated canonical pathway was "Actin Cytoskeleton Signaling", which plays a pivotal role in cell migration (Table S4 and Figure S3); this finding is consistent with the notion that some differential *O*-GlcNAcylated proteins may be involved in promoting lung cancer cell migration. Moreover, the molecular and cellular functions of these glycoproteins identified by IPA were mostly associated with cellular growth, proliferation, survival, assembly, and organization (Table S5), which generally agreed with the findings of GO analysis.

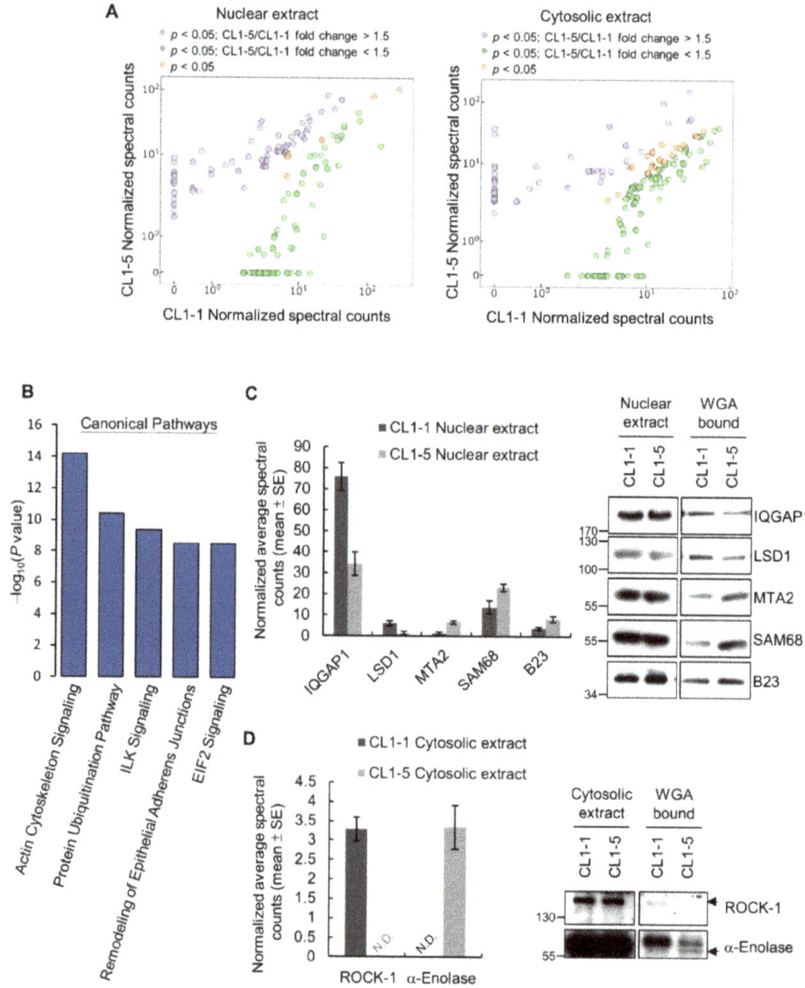

Figure 2. Differential WGA-bound glycoproteins identified in CL1-1 and CL1-5 cells. (**A**) Scatter plots of the normalized spectral counts of proteins with statistically significant (Student's *t*-test, *p*-value < 0.05) fold changes in their amounts in WGA-enriched nuclear or cytosolic fractions from CL1 and CL1-5 cells. (**B**) Top canonical pathways identified by IPA of differential WGA-bound glycoproteins. (**C–D**) Validation of representative differential WGA-bound nuclear (**C**) and cytosolic (**D**) glycoproteins by Western analysis. Shown on the left are bar graphs of average normalized spectral counts listed in Tables S1 (**C**) and S2 (**D**); N.D., not determined. Representative Western results are shown on the right. Detailed information about the Western blotting can be found in Figure S8.

Among the LC-MS/MS-identified candidates of differential *O*-GlcNAcylated proteins, we picked several that have been implicated in cancer cell migration, invasion or metastasis in the literature for validation by Western analysis. We examined the amounts of WGA-bound IQ Motif Containing GTPase Activating Protein 1 (IQGAP1), Lysine-specific histone demethylase 1 (LSD1), Metastasis Associated 1 Family, Member 2 (MTA2), Src-Associated in Mitosis 68 KDa Protein (SAM68), and Nucleophosmin (NPM) in nuclear fractions and the amounts of WGA-bound Rho-associated protein kinase 1 (ROCK-1) and α-Enolase in cytosolic fractions from CL1-1 and CL1-5 cells. For all the seven examined candidates, the Western results were consistent with the observed differences in their average normalized spectral counts in LC-MS/MS profiling (Figure 2C,D). To our knowledge, there are no reports describing the investigation of O-GlcNAc modification of any of these seven candidates. However, we found all seven of them listed as O-GlcNAcylated proteins in the O-GlcNAc Database v1.2 (https://www.oglcnac.mcw.edu, accessed on 14 Dec 2021), which is a human O-GlcNAcome database created by the Olivier-Van Stichelen lab at the Medical College of Wisconsin Department of Biochemistry using published O-GlcNAcome data [29]. In fact, out of the 264 differential WGA-bound proteins we identified, 246 (246/264; 93%) were listed as O-GlcNAcylated proteins in the O-GlcNAc Database v1.2 (Tables S1 and S2). Together, these results suggest that the profiling workflow was effective in identifying differential O-GlcNAcylated proteins in cancer cells with different degrees of invasiveness.

3.3. *SAM68 Is an O-GlcNAcylated Protein Associated with OGT in Lung Cancer Cells*

We chose SAM68 for further characterization of its O-GlcNAcylation and involvement in lung cancer aggressiveness. SAM68 is a nuclear RNA-binding protein belonging to the signal transduction and activation of RNA metabolism (STAR) family [30–32], and evidence in the literature has linked SAM68 to tumorigenesis and progression of different cancers [33–37]. Although SAM68 has been discovered as an O-GlcNAcylated protein in a high-throughput study [38], its O-GlcNAcylation remains uncharacterized. Given that a higher amount of SAM68 was pulled-down by WGA from the nuclear extract of CL1-5 cells than from that of CL1-1 cells, we sought to confirm the O-GlcNAc modification on SAM68 in lung cancer cells by immunoprecipitation of SAM68 from total cell lysates of CL1-5 and A549 cells and Western analysis using anti-O-GlcNAc antibodies. Results demonstrated that endogenous SAM68 was O-GlcNAcylated in both cell lines (Figure 3A). We explored the physical interaction between SAM68 and OGT by co-immunoprecipitation. In the samples immunoprecipitated by anti-SAM68 antibodies but not in those by IgG, a faint signal of OGT was detected; considering that the interaction might be transient and/or weak, we employed the cross-linker dithiobis [succinimidylpropionate] (DSP), and the results showed that OGT was co-immunoprecipitated by SAM68-specific antibodies from lysates of DSP-treated CL1-5 cells (Figure 3B). Reciprocal experiments using anti-OGT antibodies for immunoprecipitation detected co-immunoprecipitated SAM68 in the sample of untreated cells but not in the sample of DSP-treated cells, probably because the antibodies immunoprecipitated much less OGT in the DSP-treated sample (Figure 3C). Furthermore, we conducted immunofluorescence staining to examine the localization of SAM68 and OGT in CL1-5 cells transiently expressing Flag-tagged SAM68 or Myc-tagged OGT. The signals of SAM68 and OGT were predominantly co-localized in the nucleus (Figure 3D). Collectively, these results confirmed that SAM68 is O-GlcNAcylated and suggested that it interacts with OGT mainly in the nucleus.

3.4. *The N-Terminal Region of SAM68 Is Crucial for Its O-GlcNAcylation*

SAM68 contains a total of 47 Ser/Thr residues in its sequence and most of them are located within the N-terminal 100 aa of the protein (Figure 4A). To explore the region important for O-GlcNAcylation in SAM68, we generated a series of N-terminal truncation mutants of SAM68; results of immunoprecipitation and Western analysis of these mutants showed that deletion of aa 1–26 significantly decreased the O-GlcNAcylation of SAM68 and deletion of aa 1–56 or aa 1–95 further diminished the modification (Figure 4B), indicating the importance of the N-terminal region for SAM68 O-GlcNAcylation. We next surveyed the O-GlcNAc-modified residues in SAM68 by MS/MS analysis of immunoprecipitated SAM68 or WGA-bound proteins, and results from multiple experiments revealed 11 O-GlcNAcylation sites (Figure S4). Notably, in the N-terminal region mapped to be important for SAM68 O-GlcNAcylation, S12, S15, S18, S20, and S24 were confirmed to be O-GlcNAcylation sites. We further investigated the aa 1–26 region by preparing single-site mutants with individual Ser residues in this region replaced with Ala;

results showed that all generated single-site mutants still retained significant amounts of O-GlcNAcylation on SAM68 (Figures S5 and S15). We then prepared a multiple-site mutant 6A (S12A/S14A/S15A/S18A/S20A/S24A). When lysates from CL1-5 and 293T cells transfected with plasmids expressing wild-type (WT) or 6A mutant SAM68 were subjected to WGA pull-down and Western analysis, we found the amount of WGA-bound mutant was significantly smaller than that of WGA-bound WT SAM68 in both cell lines (Figure 4C). Additionally, SAM68 immunoprecipitation combined with O-GlcNAc Western analysis of lysates from WT or 6A SAM68-expressing CL1-5 cells confirmed that O-GlcNAcylation on 6A mutant SAM68 was significantly lower than that on WT SAM68 (Figure 4D). These data together suggest that multiple serine residues in the N-terminal 26 aa region are involved in O-GlcNAcylation of SAM68.

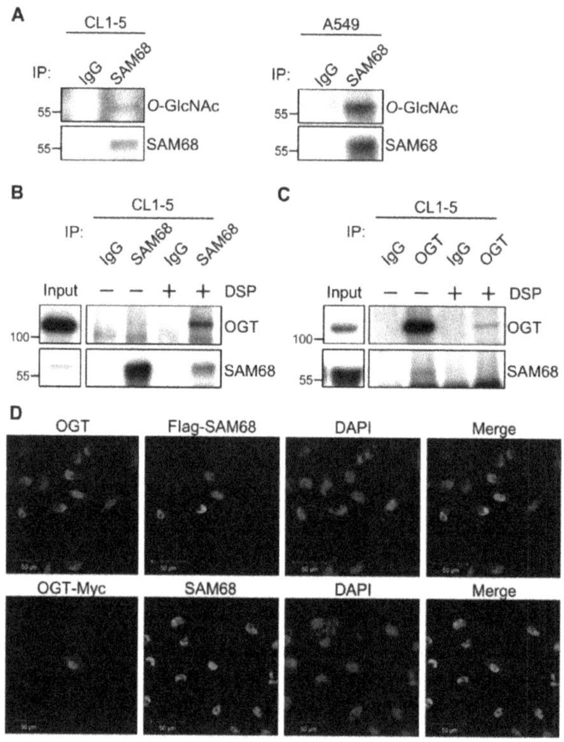

Figure 3. O-GlcNAcylation of SAM68 and its interaction with OGT. (**A**) O-GlcNAc Western analysis of immunoprecipitated SAM68. Total lysates from CL1-5 and A549 cells were subjected to immunoprecipitation with anti-SAM68 or IgG antibodies and subsequent Western blotting using the antibodies indicated. (**B**,**C**) Co-immunoprecipitation. Lysates from CL1-5 cells treated with or without the cross-linker DSP were subjected to immunoprecipitation using anti-SAM68 (**B**) or anti-OGT (**C**) antibodies, and immunoprecipitated proteins were analyzed by Western blotting with indicated antibodies. (**D**) Immunofluorescence cell staining. CL1-5 cells transiently expressing Flag-tagged SAM68 or Myc-tagged OGT were fixed and subjected to immune-staining using primary antibodies recognizing OGT, Flag, Myc, and SAM68 as indicated; DAPI was used to stain cell nuclei. Micrographs were captured under a fluorescence confocal microscope. Detailed information about the Western blotting can be found at Figure S9. OGT, O-GlcNAc transferase; DSP, dithiobis (succinimidyl propionate).

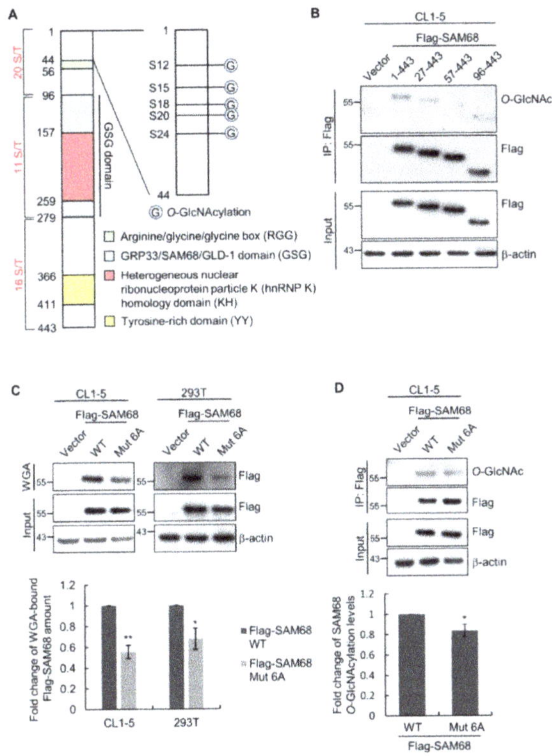

Figure 4. Characterization of SAM68 *O*-GlcNAcylation by truncation analysis and site-directed mutagenesis. (**A**) A schematic of the SAM68 domain structure. Numbers are aa positions. Total numbers of Ser/Thr (S/T) residues in indicated regions are shown. (**B**) Comparison of *O*-GlcNAcylation levels of full-length SAM68 (WT; 1-443 aa) and truncation mutants (indicated by remaining aa). Whole cell lysates from CL1-5 cells transfected with the control vector or SAM68 expression constructs were subjected to immunoprecipitation and Western analysis as indicated. (**C**,**D**) Comparison of *O*-GlcNAcylation levels of wild-type (WT) and a mutant SAM68 (Mut 6A; with S12A/S14A/S15A/S18A/S20A/S24A substitutions). CL1-5 and 293T cells were transfected for 24 hrs to express WT or Mut 6A before analysis. Representative Western results and quantitative data from multiple independent experiments are shown. (**C**) Lysates were subjected to WGA-pulldown and subsequent Western analysis. (**D**) Lysates were subjected to immunoprecipitation and subsequent Western analysis. All quantitative data shown are the means ± SD; *, $p < 0.05$; **, $p < 0.01$. Detailed information about the Western blotting can be found in Figures S10–S12.

3.5. O-GlcNAcylation Sites in the N-Terminal Region of SAM68 Are Important for Regulating Lung Cancer Cell Migration and Invasion

For investigating the role of SAM68 in lung cancer cell aggressiveness, we obtained CL1-5 clones in which the expression of SAM68 was stably suppressed by two different SAM68-targeting shRNAs (shSAM68 #1 and shSAM68 #2) (Figure 5A). Compared to the control cells, silencing of SAM68 expression markedly decreased the migration (Figure 5B) and invasion (Figure 5C) of CL1-5 cells in Transwell assays. To explore the importance of *O*-GlcNAc modifications in the *N*-terminal region of SAM68 for regulating cancer cell migration and invasion, we expressed the WT and mutant 6A of SAM68 in shSAM68 #2 cells (Figure 5D). Transwell migration assays demonstrated that shSAM68 #2 cells with reconstituted expression of WT but not mutant SAM68 showed increased migration compared to the vector control cells (Figure 5E). Similarly, reconstitution of SAM68 expression in shSAM68 #2 cells increased the invasive capability of cells, but this effect was significantly blunted if SAM68 mutant 6A was used for reconstitution (Figure 5F). Together, these findings are consistent with the notion that *O*-GlcNAcylation in the *N*-terminal region of SAM68 may promote lung cancer aggressiveness by enhancing the migratory and invasive abilities of cancer cells.

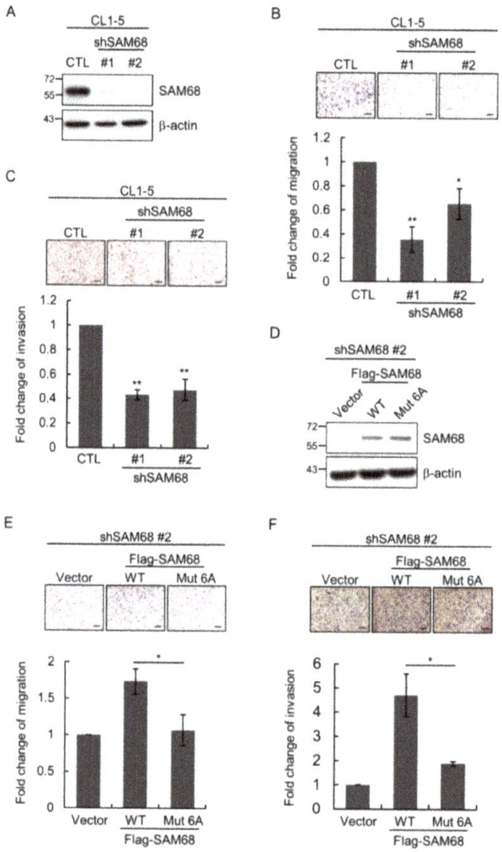

Figure 5. *O*-GlcNAcylation in the *N*-terminal region of SAM68 and its functional significance. (**A–C**) Two different *KHDRBS1*-targeting shRNAs (shSAM68 #1 and #2) were used to generate independent SAM68-knockdown CL1-5 clones. The LKO vector was used as an infection control (CTL). Cells were subjected to Western analysis (**A**), Transwell migration assays (**B**), and Transwell invasion assay (**C**). (**D–F**) CL1-5 shSAM68 #2 cells were transfected by the control vector or a construct to express Flag-tagged SAM68 WT or mutant 6A, and resulting cells were subjected to Western analysis (**D**), Transwell migration assays (**E**), and Transwell invasion assays (**F**). All quantitative data shown are the means ± SD of multiple independent experiments; *, $p < 0.05$; **, $p < 0.01$. Scale bar, 100 μm. Detailed information about the Western blotting can be found in Figure S13.

3.6. Association of SAM68 Expression with Cancer Stage and Clinical Outcome of Patients with Lung Adenocarcinoma

Clinical significance of SAM68 in lung cancer was previously suggested by studies showing upregulation of *KHDRBS1* transcript (which encodes SAM68) or SAM68 protein levels in NSCLC or lung adenocarcinoma cancerous tissues [37,39–41]; however, association with poor outcome of patients with adenocarcinoma was only shown for high *KHDRBS1* transcript levels. To explore the prognostic association at the protein level, we examined a cohort of 174 patients with lung adenocarcinoma by performing immunohistochemistry (IHC) analysis for SAM68 expression on tissue microarrays constructed using specimens from these patients. The IHC result was scored by pathologists based on the intensity (Figure 6A) and percentage of staining. When the IHC results were analyzed in relation to clinical data, we found that tissues of lung adenocarcinoma at late stages (stage II–IV) exhibited higher levels of SAM68 than those at stage I (Figure 6B and Table S6). However, when we categorized patients into low- and high-SAM68 expression subgroups based on the median IHC score and performed Kaplan-Meier analysis, the difference in the overall survival of these two subgroups did not reach statistical significance (Figure S6). Considering our above-mentioned results suggesting an important role of SAM68 *O*-GlcNAcylation in modulating cancer cell migration and invasion, we further categorized the patients according to OGT and *O*-GlcNAc IHC scores and compared the outcome between subgroups. Kaplan-Meier analysis revealed that the high-SAM68/high-OGT subgroup had shorter overall survival than the low-SAM68/low-OGT subgroup (Figure 6C, $p = 0.016$, log-rank analysis), but no significant difference was observed between the high-SAM68/high-*O*-GlcNAc and low-SAM68/low-*O*-GlcNAc subgroups (Figure 6D, $p = 0.109$).

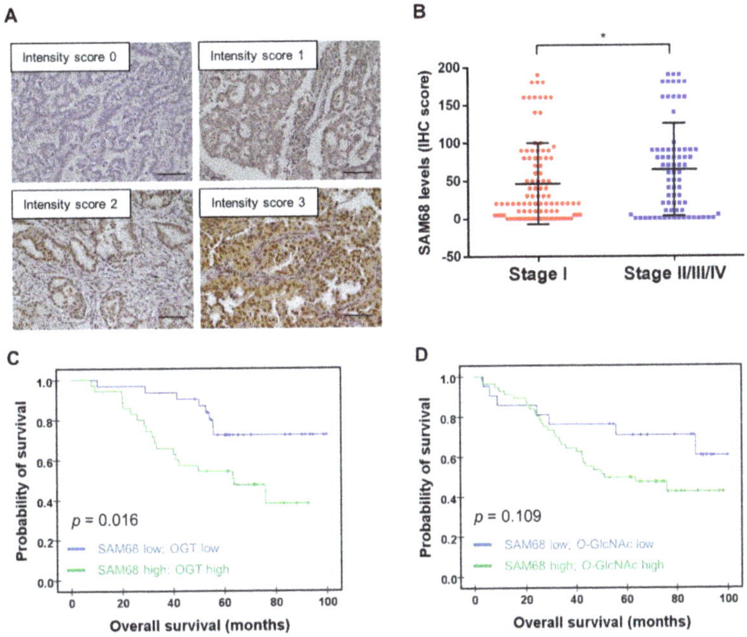

Figure 6. Association of SAM68 expression with lung cancer stage and patient outcome. (**A**) Analysis of SAM68 protein expression in clinical specimens of lung adenocarcinoma by immunohistochemistry (IHC). Shown are images representing different IHC staining intensity scores: negative (0), weak (1), moderate (2), and strong (3). Scale bar, 100 µm. (**B**) Analysis of SAM68 expression in lung adenocarcinoma tissues in relation to cancer stage. (**C–D**) Kaplan-Meier survival analysis of patients with lung adenocarcinoma categorized according to SAM68/OGT (**C**) or SAM68/*O*-GlcNAc (**D**) levels. *p*-values were derived from the log-rank test; *, $p < 0.05$.

4. Discussion

Evidence indicates that *O*-GlcNAcylation plays important roles in tumor proliferation, resistance to apoptosis, metabolism, and metastasis; many oncogenic factors or tumor suppressors have been shown to be *O*-GlcNAcylated proteins [13,42,43]. Hence, identification of cancer-associated *O*-GlcNAcylated proteins should aid in our understanding of molecular mechanisms underlying cancer pathogenesis. To date, only a handful of publications describe proteomic profiling of cancer-related O-GlcNAcylated proteins associated with breast cancer [44,45], colorectal cancer [46], and cholangiocarcinoma [47]. To our knowledge, invasion-associated O-GlcNAc proteomes of human lung cancer cells have not been explored. In this study, we used WGA enrichment and mass spectrometry to reveal O-GlcNAcylated proteins that may relate to lung adenocarcinoma aggressiveness. Among the identified candidate O-GlcNAcylated proteins, SAM68 was validated to be O-GlcNAcylated and was shown to be associated with OGT in the nucleus of lung cancer cells. Additionally, we obtained evidence suggesting that O-GlcNAcylated SAM68 promotes the migratory and invasive abilities of lung cancer cells. Furthermore, IHC analysis of clinical specimens revealed the association of simultaneous high expression of SAM68 and OGT with poor patient outcome.

Although we were able to identify differential WGA-bound proteins, the global O-GlcNAcylation level of nuclear/cytosolic proteins appeared to be generally similar between the two differentially invasive lung adenocarcinoma cell lines in Western analysis. In a study that identified O-GlcNAcylated proteins in invasive ductal breast carcinomas with or without lymph node metastasis [45], it was also noted that the O-GlcNAcylation status of individual proteins was independent of the overall O-GlcNAcylation levels in metastatic and non-metastatic invasive ductal breast carcinomas. Moreover, our previous IHC study found no association between the global O-GlcNAcylation status and the outcome of patients in lung adenocarcinoma [21]. These findings together highlight the importance of exploring changes in O-GlcNAc modification on specific proteins in understanding the molecular mechanisms of cancer progression.

Our profiling identified 158 down-regulated and 106 up-regulated putative O-GlcNAcylated proteins in the highly invasive compared to the lowly invasive lung adenocarcinoma cells. Results of Gene ontology (GO) analysis of these proteins generally reflect the previously understood functions of O-GlcNAcylated proteins. For example, the largest protein class of identified candidates in this study was "nucleic acid binding" (Figure S2C), which is well in line with the fact that OGT can localize to the nucleus and that numerous O-GlcNAcylated proteins are involved in transcriptional regulation [14,48]. GO Biological Process analysis identified "metabolic process" as the major process associated with the identified differential proteins (Figure S2B), which is consistent with the well-established role of cellular O-GlcNAcylation in metabolic regulation; as cellular O-GlcNAcylation status is influenced by nutrient availability via the hexosamine biosynthetic pathway (HBP), O-GlcNAcylation is considered a nutritional sensor and metabolic regulator [12]. Many studies have reported the role of O-GlcNAc modification in regulating cancer metabolism by modifying signaling proteins, metabolic enzymes, and transcription factors [17]. Our finding of the metabolic process as a major function associated with the invasiveness-related O-GlcNAc proteome emphasizes the scenario in which aberrant cancer metabolism affects tumor cell migration and invasion. It is established that metabolic reprogramming in cancer cells may lead to HIF-1α activation, which reduces E-cadherin expression and promotes the epithelial-mesenchymal transition and cancer cell invasion [49]. However, we did not identify HIF-1α in this study; the molecular basis linking O-GlcNAcylated protein-mediated modulation of cancer metabolism and lung cancer cell invasiveness requires further investigation. Importantly, the most significant canonical pathways associated with the invasiveness-associated O-GlcNAc proteome revealed in this study, which included actin cytoskeleton signaling, ILK signaling, and remodeling of epithelial adherens junctions as revealed by the IPA (Table S4), stress the role of O-GlcNAcylation in modulating the functions of key molecular players in these pathways. Future exploration of these O-GlcNAcylated proteins may expand our understanding of how O-GlcNAcylation regulates lung cancer cell invasion.

Our data here clearly demonstrated that the nuclear RNA-binding protein SAM68 is highly O-GlcNAcylated, with O-GlcNAc modification on multiple sites. SAM68 contains several defined domains in its protein structure [33]. The N-terminal region contains several Pro-rich motifs that may mediate the interaction with SH3 domain-containing proteins [50] and an RGG (Arginine-Glycine-Glycine) box [33]. The RNA-binding function of SAM68 is provided by an hnRNP-K homology domain (KH domain); additionally, there is a Tyr-rich (YY) domain near the C-terminus [33]. Among the 11 *O*-GlcNAcylated sites detected by MS analysis in our study, five (S12, S15, S18, S20, S24) locate in the N-terminal region but not in the Pro-rich motifs or the RGG box, two (T183, S202) are in the KH domain, three (T317, T324, T330) reside in the YY domain, and one (S422) is in the C-terminal region (Figure S4A). We noted that S18, S20, and T183 have also been identified as phosphorylated sites in a high-throughput study [51]. It will be of interest to investigate how these different types of protein modification of SAM68 functionally interact.

We focused our further investigation on the *N*-terminal region of SAM68 because analysis of truncation mutants suggested its importance for *O*-GlcNAcylation and because its role in SAM68 functions is not clear. The *N*-terminal region of SAM68 has a very high tendency of a disordered structure, as predicted by IUPred [50]. Intrinsically disordered regions are usually susceptible to diverse post-translational modifications and are characterized by marked conformational flexibility and structural plasticity to regulate protein-protein interactions [52,53]. Consistent with this notion, we identified multiple *O*-GlcNAcylated sites within this region of SAM68; future studies are obviously needed to elucidate the effect of *O*-GlcNAcylation of these sites on the interaction of SAM68 with its binding partners. Our finding that mutating *O*-GlcNAcylated sites in this region disrupt the function of SAM68 in the regulation of lung adenocarcinoma cell migration/invasion advocates a functional role of *O*-GlcNAcylation. However, as mentioned above that several identified *O*-GlcNAcylated Ser residues may also be phosphorylation sites, at this point we cannot rule out the possibility that phosphorylation may also modulate the function of SAM68 in regulating cancer cell aggressiveness. More detailed investigation is required to clarify the significance of these modifications of SAM68.

SAM68 is known to play important roles in the regulation of mRNA processing, signal transduction, gene transcription, and DNA repair [33,54], and emerging evidence links SAM68 to pathogenic mechanisms, including cell proliferation, apoptosis, invasion, and metastasis in various human cancers [55–58]. Previous studies and our data here agree in that upregulation of SAM68 expression is associated with lung cancer progression and poor patient outcome [37,39–41]. However, the molecular mechanisms in which SAM68 participates in the pathogenesis of lung cancer still await elucidation. It has been suggested that SAM68 can promote the proliferation of NSCLC cells by activating the Wnt/β-catenin pathway [40]. A recent report indicates that SAM68 promotes tumorigenesis and cancer metabolic programming in lung adenocarcinoma cells by regulating RNA splicing to increase the formation of oncogenic pyruvate kinase PKM2; the C-terminal region (aa 351-443) of SAM68 is a key functional domain in this regulation [37]. Here, we have added that *O*-GlcNAcylation of SAM68 in the *N*-terminal region has functional significance in lung cancer cell migration and invasion. Together, findings obtained by us and others establish that SAM68 plays a pivotal in lung cancer pathogenesis, and pursuing the understanding of its molecular regulation should help in guiding the development of novel SAM68-based diagnostic or therapeutic reagents.

5. Conclusions

We obtained a proteomic profile containing 264 candidates for invasiveness-related *O*-GlcNAcylated proteins in lung adenocarcinoma cells and further characterized SAM68 concerning its *O*-GlcNAcylation and involvement in cancer cell aggressiveness. Our data suggest that *O*-GlcNAcylation sites in the *N*-terminal region are important for the function of SAM68 in regulating cancer cell migration and invasion and that concomitant high expression of SAM68 and OGT in lung adenocarcinoma tissues is associated with poor patient outcome. These findings highlight the potential of invasiveness-associated *O*-GlcNAcylated proteins as novel biomarkers for lung cancer prognosis. Further understanding of the functional consequence of differential *O*-GlcNAcylation of proteins identified in this study may provide new insights into the mechanisms of lung cancer progression.

Supplementary Materials: The following are available online at https://www.mdpi.com/article/10.3390/cancers14010243/s1, Figure S1: Enrichment of nuclear/cytosolic *O*-GlcNAcylated proteins, Figure S2: Gene ontology analysis, Figure S3: The actin cytoskeleton signaling pathway, Figure S4: Identification of *O*-GlcNAcylation sites by MS/MS analysis, Figure S5: Comparison of *O*-GlcNAcylation levels of wild-type and mutant SAM68, Figure S6: Kaplan-Meier analysis of the overall survival of patients with lung adenocarcinoma, Figure S7: Full Western blot for Figure 1B, Figure S8: Full Western blot for Figure 2C,D, Figure S9: Full Western blot for Figure 3A–C, Figure S10: Full Western blot for Figure 4B, Figure S11: Full Western blot for Figure 4C, Figure S12: Full Western blot for Figure 4D, Figure S13: Full Western blot for Figure 5A and 5D, Figure S14: Full Western blot for Figure S1, Figure S15: Full Western blot for Figure S5, Table S1: Differential nuclear WGA-bound proteins, Table S2: Differential cytosolic WGA-bound proteins, Table S3: Significant network functions, Table S4: Significant canonical pathways, Table S5: Significant molecular and cellular functions, Table S6: Association of SAM68 expression with lung adenocarcinoma stage.

Author Contributions: Conceptualization, T.-Y.C. and M.-Y.C.; data curation, C.-H.L., Y.-C.Y. and C.-C.L.; formal analysis, C.-H.L., S.-Y.W., C.-Y.P. and C.-C.L.; funding acquisition, T.-Y.C. and M.-Y.C.; investigation, C.-H.L., S.-Y.W., C.-Y.P. and C.-C.L.; methodology, C.-C.L.; project administration, C.-H.L., C.-C.L., M.-Y.C. and T.-Y.C.; resources, C.-C.L., Y.-C.Y. and T.-Y.C.; software, C.-C.L.; supervision, C.-H.L., C.-C.L., M.-Y.C. and T.-Y.C.; validation, C.-H.L., S.-Y.W. and C.-Y.P.; visualization, C.-H.L., S.-Y.W., C.-Y.P., Y.-C.Y. and C.-C.L.; writing—original draft preparation, C.-H.L. and M.-Y.C.; writing—review and editing, C.-C.L., M.-Y.C. and T.-Y.C. All authors have read and agreed to the published version of the manuscript.

Funding: This research was supported by the Ministry of Science and Technology, Taiwan (Grant No.: MOST105-2320-B-075-003-MY3 and MOST108-2320-B-075-001-MY3) and the "Cancer Progression Research Center, National Yang Ming Chaio Tung University" from The Featured Areas Research Center Program within the framework of the Higher Education Sprout Project by the Ministry of Education, Taiwan.

Institutional Review Board Statement: The study was conducted according to the guidelines of the Declaration of Helsinki and was approved by the Institutional Review Board of the Taipei Veterans General Hospital, Taiwan (IRB No. 2021-04-011BCF).

Informed Consent Statement: Patient consent was waived as the human samples used in this study were from de-identified residual specimens of formalin-fixed, paraffin-embedded tissues for surgical pathology examination.

Data Availability Statement: The data presented in this report are available from the corresponding authors upon reasonable request.

Acknowledgments: We thank Pan-Chyr Yang from National Taiwan University, Taiwan, for providing cell lines and Neng-Yao Shih from the National Health Research Institutes, Taiwan, for providing antibodies. We acknowledge the RNA Technology Platform and Gene Manipulation Core, Academia Sinica, for reagents to perform lentiviral shRNA-mediated gene silencing.

Conflicts of Interest: The authors declare no conflict of interest.

References

1. Sung, H.; Ferlay, J.; Siegel, R.L.; Laversanne, M.; Soerjomataram, I.; Jemal, A.; Bray, F. Global Cancer Statistics 2020: GLOBOCAN Estimates of Incidence and Mortality Worldwide for 36 Cancers in 185 Countries. *CA A Cancer J. Clin.* **2021**, *71*, 209–249. [CrossRef]
2. Novello, S.; Barlesi, F.; Califano, R.; Cufer, T.; Ekman, S.; Levra, M.G.; Kerr, K.; Popat, S.; Reck, M.; Senan, S.; et al. Metastatic non-small-cell lung cancer: ESMO Clinical Practice Guidelines for diagnosis, treatment and follow-up. *Ann. Oncol. Off. J. Eur. Soc. Med Oncol.* **2016**, *27*, v1–v27. [CrossRef]
3. Matsuda, T.; Machii, R. Morphological distribution of lung cancer from Cancer Incidence in Five Continents Vol. X. *Jpn. J. Clin. Oncol.* **2015**, *45*, 404. [CrossRef]
4. Wang, B.Y.; Huang, J.Y.; Cheng, C.Y.; Lin, C.H.; Ko, J.; Liaw, Y.P. Lung cancer and prognosis in taiwan: A population-based cancer registry. *J. Thorac. Oncol. Off. Publ. Int. Assoc. Study Lung Cancer* **2013**, *8*, 1128–1135. [CrossRef]
5. Lambert, A.W.; Pattabiraman, D.R.; Weinberg, R.A. Emerging Biological Principles of Metastasis. *Cell* **2017**, *168*, 670–691. [CrossRef] [PubMed]
6. Torres, C.R.; Hart, G.W. Topography and polypeptide distribution of terminal N-acetylglucosamine residues on the surfaces of intact lymphocytes. Evidence for O-linked GlcNAc. *J. Biol. Chem.* **1984**, *259*, 3308–3317. [CrossRef]

7. Haltiwanger, R.S.; Holt, G.D.; Hart, G.W. Enzymatic addition of O-GlcNAc to nuclear and cytoplasmic proteins. Identification of a uridine diphospho-N-acetylglucosamine:peptide beta-N-acetylglucosaminyltransferase. *J. Biol. Chem.* **1990**, *265*, 2563–2568. [CrossRef]
8. Haltiwanger, R.S.; Blomberg, M.A.; Hart, G.W. Glycosylation of nuclear and cytoplasmic proteins. Purification and characterization of a uridine diphospho-N-acetylglucosamine:polypeptide beta-N-acetylglucosaminyltransferase. *J. Biol. Chem.* **1992**, *267*, 9005–9013. [CrossRef]
9. Dong, D.L.; Hart, G.W. Purification and characterization of an O-GlcNAc selective N-acetyl-beta-D-glucosaminidase from rat spleen cytosol. *J. Biol. Chem.* **1994**, *269*, 19321–19330. [CrossRef]
10. Gao, Y.; Wells, L.; Comer, F.I.; Parker, G.J.; Hart, G.W. Dynamic O-glycosylation of nuclear and cytosolic proteins: Cloning and characterization of a neutral, cytosolic beta-N-acetylglucosaminidase from human brain. *J. Biol. Chem.* **2001**, *276*, 9838–9845. [CrossRef] [PubMed]
11. Chatham, J.C.; Zhang, J.; Wende, A.R. Role of O-Linked N-Acetylglucosamine Protein Modification in Cellular (Patho)Physiology. *Physiol Rev.* **2021**, *101*, 427–493. [CrossRef] [PubMed]
12. Hart, G.W. Nutrient regulation of signaling and transcription. *J. Biol. Chem.* **2019**, *294*, 2211–2231. [CrossRef] [PubMed]
13. Hanover, J.A.; Chen, W.; Bond, M.R. O-GlcNAc in cancer: An Oncometabolism-fueled vicious cycle. *J. Bioenerg Biomembr.* **2018**, *50*, 155–173. [CrossRef]
14. Parker, M.P.; Peterson, K.R.; Slawson, C. O-GlcNAcylation and O-GlcNAc Cycling Regulate Gene Transcription: Emerging Roles in Cancer. *Cancers* **2021**, *13*, 1666. [CrossRef] [PubMed]
15. Zhu, Y.; Hart, G.W. Targeting O-GlcNAcylation to develop novel therapeutics. *Mol. Asp. Med.* **2021**, *79*, 100885. [CrossRef]
16. Zhu, Y.; Shan, X.; Yuzwa, S.A.; Vocadlo, D.J. The emerging link between O-GlcNAc and Alzheimer disease. *J. Biol. Chem.* **2014**, *289*, 34472–34481. [CrossRef] [PubMed]
17. Ma, Z.; Vosseller, K. Cancer metabolism and elevated O-GlcNAc in oncogenic signaling. *J. Biol. Chem.* **2014**, *289*, 34457–34465. [CrossRef]
18. Park, S.Y.; Kim, H.S.; Kim, N.H.; Ji, S.; Cha, S.Y.; Kang, J.G.; Ota, I.; Shimada, K.; Konishi, N.; Nam, H.W.; et al. Snail1 is stabilized by O-GlcNAc modification in hyperglycaemic condition. *EMBO J.* **2010**, *29*, 3787–3796. [CrossRef]
19. Huang, X.; Pan, Q.; Sun, D.; Chen, W.; Shen, A.; Huang, M.; Ding, J.; Geng, M. O-GlcNAcylation of cofilin promotes breast cancer cell invasion. *J. Biol. Chem.* **2013**, *288*, 36418–36425. [CrossRef] [PubMed]
20. Tao, T.; He, Z.; Shao, Z.; Lu, H. TAB3 O-GlcNAcylation promotes metastasis of triple negative breast cancer. *Oncotarget* **2016**, *7*, 22807–22818. [CrossRef]
21. Lin, Y.C.; Lin, C.H.; Yeh, Y.C.; Ho, H.L.; Wu, Y.C.; Chen, M.Y.; Chou, T.Y. High O-linked N-acetylglucosamine transferase expression predicts poor survival in patients with early stage lung adenocarcinoma. *Oncotarget* **2018**, *9*, 31032–31044. [CrossRef]
22. Mi, W.; Gu, Y.; Han, C.; Liu, H.; Fan, Q.; Zhang, X.; Cong, Q.; Yu, W. O-GlcNAcylation is a novel regulator of lung and colon cancer malignancy. *Biochim. Et Biophys. Acta* **2011**, *1812*, 514–519. [CrossRef] [PubMed]
23. Lucena, M.C.; Carvalho-Cruz, P.; Donadio, J.L.; Oliveira, I.A.; de Queiroz, R.M.; Marinho-Carvalho, M.M.; Sola-Penna, M.; de Paula, I.F.; Gondim, K.C.; McComb, M.E.; et al. Epithelial Mesenchymal Transition Induces Aberrant Glycosylation through Hexosamine Biosynthetic Pathway Activation. *J. Biol. Chem.* **2016**, *291*, 12917–12929. [CrossRef]
24. Perez-Riverol, Y.; Csordas, A.; Bai, J.; Bernal-Llinares, M.; Hewapathirana, S.; Kundu, D.J.; Inuganti, A.; Griss, J.; Mayer, G.; Eisenacher, M.; et al. The PRIDE database and related tools and resources in 2019: Improving support for quantification data. *Nucleic Acids Res.* **2019**, *47*, D442–D450. [CrossRef]
25. Liao, C.; Chang, S.; Hu, S.; Tang, Z.; Fu, G. Rapid and sensitive liquid chromatography-tandem mass spectrometry method for determination of 1-beta-d-arabinofuranosyluracil in human plasma and application to therapeutic drug monitoring in patient with leukemia. *J. Pharm. Biomed. Anal.* **2013**, *85*, 118–122. [CrossRef]
26. Mi, H.; Muruganujan, A.; Ebert, D.; Huang, X.; Thomas, P.D. PANTHER version 14: More genomes, a new PANTHER GO-slim and improvements in enrichment analysis tools. *Nucleic Acids Res.* **2019**, *47*, D419–D426. [CrossRef] [PubMed]
27. Lin, S.C.; Lin, C.H.; Shih, N.C.; Liu, H.L.; Wang, W.C.; Lin, K.Y.; Liu, Z.Y.; Tseng, Y.J.; Chang, H.K.; Lin, Y.C.; et al. Cellular prion protein transcriptionally regulated by NFIL3 enhances lung cancer cell lamellipodium formation and migration through JNK signaling. *Oncogene* **2019**, *39*, 385–398. [CrossRef]
28. Chu, Y.W.; Yang, P.C.; Yang, S.C.; Shyu, Y.C.; Hendrix, M.J.; Wu, R.; Wu, C.W. Selection of invasive and metastatic subpopulations from a human lung adenocarcinoma cell line. *Am. J. Respir. Cell Mol. Biol.* **1997**, *17*, 353–360. [CrossRef]
29. Wulff-Fuentes, E.; Berendt, R.R.; Massman, L.; Danner, L.; Malard, F.; Vora, J.; Kahsay, R.; Olivier-Van Stichelen, S. The human O-GlcNAcome database and meta-analysis. *Sci. Data* **2021**, *8*, 25. [CrossRef]
30. Lukong, K.E.; Richard, S. Sam68, the KH domain-containing superSTAR. *Biochim. Et Biophys. Acta* **2003**, *1653*, 73–86. [CrossRef] [PubMed]
31. Taylor, S.J.; Shalloway, D. An RNA-binding protein associated with Src through its SH2 and SH3 domains in mitosis. *Nature* **1994**, *368*, 867–871. [CrossRef] [PubMed]
32. Fumagalli, S.; Totty, N.F.; Hsuan, J.J.; Courtneidge, S.A. A target for Src in mitosis. *Nature* **1994**, *368*, 871–874. [CrossRef] [PubMed]
33. Frisone, P.; Pradella, D.; Di Matteo, A.; Belloni, E.; Ghigna, C.; Paronetto, M.P. SAM68: Signal Transduction and RNA Metabolism in Human Cancer. *BioMed Res. Int.* **2015**, *2015*, 528954. [CrossRef]
34. Fu, K.; Sun, X.; Wier, E.M.; Hodgson, A.; Liu, Y.; Sears, C.L.; Wan, F. Sam68/KHDRBS1 is critical for colon tumorigenesis by regulating genotoxic stress-induced NF-kappaB activation. *eLife* **2016**, *5*. [CrossRef]

35. Li, Z.; Yu, C.P.; Zhong, Y.; Liu, T.J.; Huang, Q.D.; Zhao, X.H.; Huang, H.; Tu, H.; Jiang, S.; Zhang, Y.; et al. Sam68 expression and cytoplasmic localization is correlated with lymph node metastasis as well as prognosis in patients with early-stage cervical cancer. *Ann. Oncol. Off. J. Eur. Soc. Med Oncol.* **2012**, *23*, 638–646. [CrossRef] [PubMed]
36. Richard, S.; Vogel, G.; Huot, M.E.; Guo, T.; Muller, W.J.; Lukong, K.E. Sam68 haploinsufficiency delays onset of mammary tumorigenesis and metastasis. *Oncogene* **2008**, *27*, 548–556. [CrossRef]
37. Zhu, S.; Chen, W.; Wang, J.; Qi, L.; Pan, H.; Feng, Z.; Tian, D. SAM68 promotes tumorigenesis in lung adenocarcinoma by regulating metabolic conversion via PKM alternative splicing. *Theranostics* **2021**, *11*, 3359–3375. [CrossRef] [PubMed]
38. Nandi, A.; Sprung, R.; Barma, D.K.; Zhao, Y.; Kim, S.C.; Falck, J.R.; Zhao, Y. Global identification of O-GlcNAc-modified proteins. *Anal. Chem.* **2006**, *78*, 452–458. [CrossRef]
39. Zhang, Z.; Xu, Y.; Sun, N.; Zhang, M.; Xie, J.; Jiang, Z. High Sam68 expression predicts poor prognosis in non-small cell lung cancer. *Clin. Transl. Oncol.* **2014**, *16*, 886–891. [CrossRef] [PubMed]
40. Li, X.; Zhou, X.; Hua, F.; Fan, Y.; Zu, L.; Wang, Y.; Shen, W.; Pan, H.; Zhou, Q. The RNA-binding protein Sam68 is critical for non-small cell lung cancer cell proliferation by regulating Wnt/beta-catenin pathway. *Int J. Clin. Exp. Pathol.* **2017**, *10*, 8281–8291.
41. Sumithra, B.; Saxena, U.; Das, A.B. A comprehensive study on genome-wide coexpression network of KHDRBS1/Sam68 reveals its cancer and patient-specific association. *Sci. Rep.* **2019**, *9*, 11083. [CrossRef]
42. Ferrer, C.M.; Sodi, V.L.; Reginato, M.J. O-GlcNAcylation in Cancer Biology: Linking Metabolism and Signaling. *J. Mol. Biol.* **2016**, *428*, 3282–3294. [CrossRef]
43. Makwana, V.; Ryan, P.; Patel, B.; Dukie, S.A.; Rudrawar, S. Essential role of O-GlcNAcylation in stabilization of oncogenic factors. *Biochim Biophys. Acta. Gen. Subj.* **2019**, *1863*, 1302–1317. [CrossRef]
44. Champattanachai, V.; Netsirisawan, P.; Chaiyawat, P.; Phueaouan, T.; Charoenwattanasatien, R.; Chokchaichamnankit, D.; Punyarit, P.; Srisomsap, C.; Svasti, J. Proteomic analysis and abrogated expression of O-GlcNAcylated proteins associated with primary breast cancer. *Proteomics* **2013**, *13*, 2088–2099. [CrossRef] [PubMed]
45. Jiang, K.; Gao, Y.; Hou, W.; Tian, F.; Ying, W.; Li, L.; Bai, B.; Hou, G.; Wang, P.G.; Zhang, L. Proteomic analysis of O-GlcNAcylated proteins in invasive ductal breast carcinomas with and without lymph node metastasis. *Amino Acids* **2016**, *48*, 365–374. [CrossRef] [PubMed]
46. Phueaouan, T.; Chaiyawat, P.; Netsirisawan, P.; Chokchaichamnankit, D.; Punyarit, P.; Srisomsap, C.; Svasti, J.; Champattanachai, V. Aberrant O-GlcNAc-modified proteins expressed in primary colorectal cancer. *Oncol. Rep.* **2013**, *30*, 2929–2936. [CrossRef]
47. Phoomak, C.; Park, D.; Silsirivanit, A.; Sawanyawisuth, K.; Vaeteewoottacharn, K.; Detarya, M.; Wongkham, C.; Lebrilla, C.B.; Wongkham, S. O-GlcNAc-induced nuclear translocation of hnRNP-K is associated with progression and metastasis of cholangiocarcinoma. *Mol. Oncol.* **2019**, *13*, 338–357. [CrossRef] [PubMed]
48. Seo, H.G.; Kim, H.B.; Kang, M.J.; Ryum, J.H.; Yi, E.C.; Cho, J.W. Identification of the nuclear localisation signal of O-GlcNAc transferase and its nuclear import regulation. *Sci. Rep.* **2016**, *6*, 34614. [CrossRef]
49. Kang, H.; Kim, H.; Lee, S.; Youn, H.; Youn, B. Role of Metabolic Reprogramming in Epithelial(-)Mesenchymal Transition (EMT). *Int. J. Mol. Sci.* **2019**, *20*, 2042. [CrossRef]
50. Asbach, B.; Ludwig, C.; Saksela, K.; Wagner, R. Comprehensive analysis of interactions between the Src-associated protein in mitosis of 68 kDa and the human Src-homology 3 proteome. *PLoS ONE* **2012**, *7*, e38540. [CrossRef]
51. Zhou, H.; Di Palma, S.; Preisinger, C.; Peng, M.; Polat, A.N.; Heck, A.J.; Mohammed, S. Toward a comprehensive characterization of a human cancer cell phosphoproteome. *J. Proteome Res.* **2013**, *12*, 260–271. [CrossRef]
52. Wright, P.E.; Dyson, H.J. Intrinsically disordered proteins in cellular signalling and regulation. *Nat. Rev. Mol. Cell Biol.* **2015**, *16*, 18–29. [CrossRef]
53. Owen, I.; Shewmaker, F. The Role of Post-Translational Modifications in the Phase Transitions of Intrinsically Disordered Proteins. *Int. J. Mol. Sci.* **2019**, *20*, 5501. [CrossRef] [PubMed]
54. Sun, X.; Fu, K.; Hodgson, A.; Wier, E.M.; Wen, M.G.; Kamenyeva, O.; Xia, X.; Koo, L.Y.; Wan, F. Sam68 Is Required for DNA Damage Responses via Regulating Poly(ADP-ribosyl)ation. *PLoS Biol.* **2016**, *14*, e1002543. [CrossRef] [PubMed]
55. Fu, K.; Sun, X.; Xia, X.; Hobbs, R.P.; Guo, Y.; Coulombe, P.A.; Wan, F. Sam68 is required for the growth and survival of nonmelanoma skin cancer. *Cancer Med.* **2019**, *8*, 6106–6113. [CrossRef]
56. Chen, X.; Zhang, L.; Yuan, M.; Kuang, Z.; Zou, Y.; Tang, T.; Zhang, W.; Hu, X.; Xia, T.; Cao, T.; et al. Sam68 Promotes the Progression of Human Breast Cancer through inducing Activation of EphA3. *Curr. Cancer Drug Targets* **2020**, *20*, 76–83. [CrossRef] [PubMed]
57. Wang, Q.; Li, Y.; Cheng, J.; Chen, L.; Xu, H.; Li, Q.; Pang, T. Sam68 affects cell proliferation and apoptosis of human adult T-acute lymphoblastic leukemia cells via AKT/mTOR signal pathway. *Leuk. Res.* **2016**, *46*, 1–9. [CrossRef]
58. Xiao, J.; Wang, Q.; Yang, Q.; Wang, H.; Qiang, F.; He, S.; Cai, J.; Yang, L.; Wang, Y. Clinical significance and effect of Sam68 expression in gastric cancer. *Oncol. Lett.* **2018**, *15*, 4745–4752. [CrossRef]

cancers

MDPI

Article

O-GlcNAcylation Links Nutrition to the Epigenetic Downregulation of *UNC5A* during Colon Carcinogenesis

Amélie Decourcelle [1], Ninon Very [2], Madjid Djouina [3], Ingrid Loison [1], Julien Thévenet [4], Mathilde Body-Malapel [3], Eric Lelièvre [5], Olivier Coqueret [5], Dominique Leprince [1], Ikram El Yazidi-Belkoura [2] and Vanessa Dehennaut [1,*]

1 Université de Lille, CNRS, Inserm, CHU Lille, UMR9020-U1277—CANTHER—Cancer Heterogeneity, Plasticity and Resistance to Therapies, F-59000 Lille, France; amelie.decourcelle@ibl.cnrs.fr (A.D.); ingrid.loison@ibl.cnrs.fr (I.L.); dominique.leprince@ibl.cnrs.fr (D.L.)

2 Université de Lille, CNRS, UMR 8576—UGSF—Unité de Glycobiologie Structurale et Fonctionnelle, F-59000 Lille, France; ninon.very.etu@univ-lille.fr (N.V.); ikram.el-yazidi@univ-lille.fr (I.E.Y.-B.)

3 Université de Lille, Inserm, CHU Lille, U1286—INFINITE—Institute for translational research in inflammation, F-59000 Lille, France; madjid.djouina@univ-lille.fr (M.D.); mathilde.body@univ-lille.fr (M.B.-M.)

4 Université de Lille, Inserm, CHU Lille, UMR 1190 Translational Research for Diabetes, European Genomic Institute for Diabetes, F-59000 Lille, France; julien.thevenet@univ-lille.fr

5 Paul Papin ICO Cancer Center, CRCINA, INSERM, Université de Nantes, Université d'Angers, 49055 Angers, France; eric.lelievre@univ-angers.fr (E.L.); olivier.coqueret@univ-angers.fr (O.C.)

* Correspondence: vanessa.dehennaut@ibl.cnrs.fr

Received: 2 October 2020; Accepted: 25 October 2020; Published: 28 October 2020

Simple Summary: Nutritional disorders represent major risk factors for colorectal cancer according to mechanisms of action that are still insufficiently known. The aim of our study was to investigate the putative involvement of nutrition in the epigenetic downregulation of the tumor suppressor genes of the *UNC5* (Uncoordinated 5) family during colonic carcinogenesis and to understand its molecular relays. Herein, we provided evidence that the consumption of a High Carbohydrate Diet worsens colon carcinogenesis in mice and is correlated with the downregulation of several members of the *UNC5* family whose *UNC5A* (Uncoordinated 5A). Mechanistically, we identified the nutritional sensor *O*-GlcNAcylation as one of the molecular relays that regulate the recruitment of the PRC2 complex onto the *UNC5A* promoter to repress its transcription.

Abstract: While it is now accepted that nutrition can influence the epigenetic modifications occurring in colorectal cancer (CRC), the underlying mechanisms are not fully understood. Among the tumor suppressor genes frequently epigenetically downregulated in CRC, the four related genes of the *UNC5* family: *UNC5A*, *UNC5B*, *UNC5C* and *UNC5D* encode dependence receptors that regulate the apoptosis/survival balance. Herein, in a mouse model of CRC, we found that the expression of *UNC5A*, *UNC5B* and *UNC5C* was diminished in tumors but only in mice subjected to a High Carbohydrate Diet (HCD) thus linking nutrition to their repression in CRC. *O*-GlcNAcylation is a nutritional sensor which has enhanced levels in CRC and regulates many cellular processes amongst epigenetics. We then investigated the putative involvement of *O*-GlcNAcylation in the epigenetic downregulation of the *UNC5* family members. By a combination of pharmacological inhibition and RNA interference approaches coupled to RT-qPCR (Reverse Transcription-quantitative Polymerase Chain Reaction) analyses, promoter luciferase assay and CUT&RUN (Cleavage Under Target & Release Using Nuclease) experiments, we demonstrated that the *O*-GlcNAcylated form of the histone methyl transferase EZH2 (Enhancer of Zeste Homolog 2) represses the transcription of *UNC5A* in human colon cancer cells. Collectively, our data support the hypothesis that *O*-GlcNAcylation could represent one link between nutrition and epigenetic downregulation of key tumor suppressor genes governing colon carcinogenesis including *UNC5A*.

Keywords: UNC5A; *O*-GlcNAcylation; OGT; EZH2; nutrition; epigenetics; colon cancer

1. Introduction

The emergence and progression of cancer depend on a complex interplay between the genome and the epigenome, which together interact with several environmental factors. Particularly, several studies have highlighted the key role of diet and nutritional compounds in the epigenetic regulation of gene expression [1–3]. They hence support the hypothesis of a close relationship between nutritional disorders (obesity, metabolic syndrome, type 2 diabetes, etc.), well-known risk factors for many cancers including colorectal cancer (CRC), and epigenetic reprogramming linked to carcinogenesis.

The *UNC5* gene family consists of four related genes including *UNC5A*, *UNC5B*, *UNC5C* and *UNC5D* that encode type-I transmembrane receptors of Netrin-1. UNC5 and Netrin-1 play essential role in axon guidance during neuronal development and differentiation [4]. In addition, in neuronal and non-neuronal cells, UNC5 receptors share the capability to act as "dependence receptors": they transduce a "positive" cell proliferation and survival signal when bound to Netrin-1 but induce caspase-dependent apoptosis in absence of their ligand. In recent years, the UNC5 receptors have been defined as key players of colorectal carcinogenesis by regulating the survival/apoptosis balance and are considered as conditional tumor suppressor genes [5]. In fact, expression of *UNC5A*, *UNC5B* and *UNC5C* is frequently downregulated in colorectal cancer (CRC) and their silencing has been associated in part with loss of heterozygoty (LOH) within *UNC5* loci and with epigenetic alterations that are not fully understood [6–10]. Notably, the putative influence of nutrition on the repression of the *UNC5* family members during colon carcinogenesis has not yet been investigated.

Among the molecular elements that could connect nutrition to epigenetic reprogramming in CRC, the nutritional sensor *O*-linked-β-N-acetylglucosaminylation (*O*-GlcNAcylation) has emerged during the last decade as a key regulator of chromatin remodeling and thereby of the epigenetic regulation of gene expression [11–14]. *O*-GlcNAcylation is a reversible post-translational modification of nucleocytoplasmic and mitochondrial proteins that consists in the covalent linkage of a unique sugar N-acetylglucosamine (GlcNAc) to serines and threonines of target proteins [15]. *O*-GlcNAcylation levels are regulated by a unique couple of enzymes: OGT (*O*-GlcNAc Transferase) that catalyzes the transfer of GlcNAc from UDP-GlcNAc onto the protein and OGA (*O*-GlcNAcase) that hydrolyzes the residue. *O*-GlcNAcylation levels are closely dependent upon the concentration of UDP-GlcNAc synthetized through the Hexosamine Biosynthesis Pathway (HBP) at the crossroad of glucose, amino acid, fatty acid and nucleotide metabolisms. UDP-GlcNAc and *O*-GlcNAcylation are thus considered as sensors of the nutritional state which can relay the effects of an excessive food supply, unbalanced diet, obesity and other metabolic problems that represent high risk factors of CRC [16–18]. In this sense, through Western Blot analyses, we previously showed that colons from mice fed a High Carbohydrate Diet (HCD) exhibited higher amounts of *O*-GlcNAcylation relative to mice fed a Normal Diet (ND) [19]. Moreover, we and others observed increased contents of *O*-GlcNAcylation and OGT in human colon cancer samples in comparison with normal tissues [19–23]. Contrary, decreasing *O*-GlcNAcylation levels by silencing OGT reduces proliferation, adhesion, migration and anchorage-independent cell growth of colon cancer cell lines [20,24]. Aberrant OGT and *O*-GlcNAcylation levels are thus defined as new CRC hallmarks [25].

O-GlcNAcylation is involved in the regulation of many cellular processes, including the epigenetic regulation of gene expression. Indeed, *O*-GlcNAcylation is part of the histone code and OGT interacts with and regulates the DNA demethylases of the Ten-Eleven-Translocation (TET) family and several histone modifying proteins [11–13]. Among this last class of proteins, the members of the Polycomb Repressive Complex 2 (PRC2) repress the transcription of numerous target genes through the deposit of the repressive chromatin mark H3K27Me3 consisting of trimethylation of lysine 27 on histone H3. The core of PRC2 is composed of three Polycomb group (PcG) proteins: Enhancer of Zeste Homolog 2

(EZH2), Embryonic Ectoderm Development (EED) and Suppressor of Zeste 12 (SUZ12). The methyl transferase EZH2 is the catalytic subunit of the PRC2 complex and SUZ12 and EED are indispensable for EZH2 enzymatic activity. Several studies have reported abnormally elevated expression of EZH2, EED and SUZ12 in CRC in correlation with advanced stages of the disease and poor prognosis [26–28]. Regulation of EZH2 functions by its O-GlcNAcylation has been evidenced in several studies conducted in different cell lines including breast and colon cancer cells [29–33]. In the colon cancer cell line HT29, O-GlcNAcylated proteins and H3K27 trimethylation were found together at the promoter region of 61 genes [34]. In breast cancer MCF7 cells, a co-regulation by OGT and EZH2 was also evidenced for 16 tumor suppressor genes including *UNC5A* [29]. However, the involvement of this OGT-EZH2 axis in the regulation of the expression of *UNC5A* as well as the other members of the *UNC5A* family in colon cancer cells has not been studied.

Therefore, in this study, we investigated whether nutrition could influence the expression of the *UNC5* family members during colon carcinogenesis and whether it could be related to the OGT-EZH2 axis.

2. Results

2.1. Subjecting Mice to a High Carbohydrate Diet (HCD) Worsens Colon Carcinogenesis

To test whether nutrition could be involved in the epigenetic downregulation of *UNC5* receptors during colon carcinogenesis, we subjected C57BL/6JRj mice either to a Normal Diet (ND) or to a High Carbohydrate Diet (HCD). Thirty-nine days after the beginning of the different diets, we induced CRC in these mice using the well-characterized azoxymethane (AOM)/dextran sulfate sodium (DSS) method [35] (Figure 1A). At the end of experiment, mice treated with AOM/DSS and fed HCD had a statistically significant higher blood glucose level compared to mice treated with AOM/DSS and fed ND (Supplementary Figure S1A). Moreover, weight loss was observed in mice treated with AOM/DSS and fed HCD (Supplementary Figure S1B) probably due to the severity of the disease in this group of animals. Indeed, we monitored tumor burden via endoscopy (Figure 1B) and observed that mice fed HCD had a higher number of tumors than the control group (Figure 1C) with a higher number of grade 5 tumors (Figure 1D) observed in 100% of mice (Figure 1E). We also studied the expression of *c-myc* and *cyclin D1*, two well-known target genes of the Wnt/β-catenin pathway, whose activity is commonly upregulated in CRC. As expected, both *c-myc* and *Cyclin D1* transcripts were increased in tumors in mice fed ND compared to the control group (Figure 1F, compare ND vs. ND + AOM/DSS). Interestingly, the High Carbohydrate Diet caused an even greater increase in *c-myc* and *Cyclin D1* expression (Figure 1F, compare ND + AOM/DSS vs. HCD + AOM/DSS). Moreover, in mice treated with AOM/DSS in conjunction with the HCD, we also observed a clear decrease in colon length compared to mice fed ND (Figure 1G,H) and an increase in *Cox-2* and *IκBα* mRNA levels (that indicates activation of the NF-κB pathway) (Figure 1F) thus reflecting a higher level of inflammation in these mice. Taken together, these results demonstrate, as we might expect, that the High Carbohydrate Diet worsens inflammation-driven colon carcinogenesis in mice.

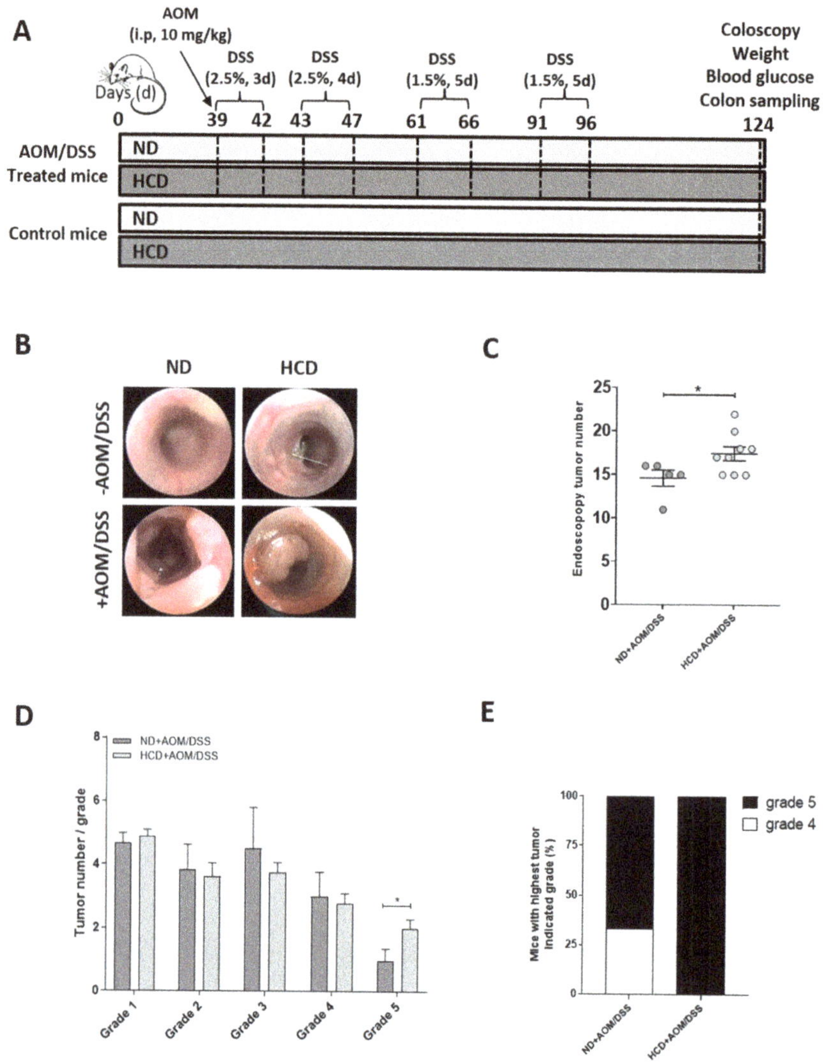

Figure 1. *Cont.*

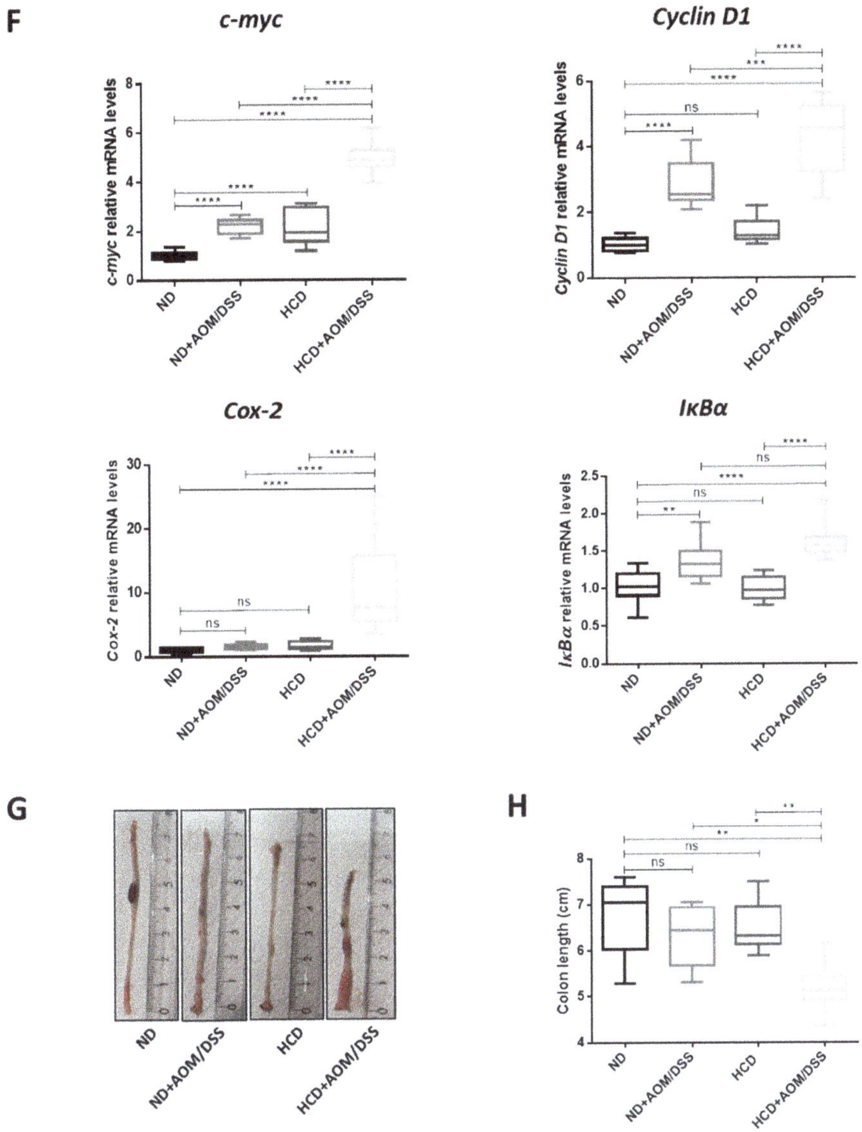

Figure 1. Subjecting mice to a High Carbohydrate Diet (HCD) worsens colon carcinogenesis. (**A**) Experimental protocol. Half of the C57BL/6JRj mice were subjected to a Normal Diet (ND) and the other half to a High Carbohydrate Diet (HCD). Thirty-nine days after the beginning of the two diets, half of the experimental animals received a single intraperitoneal injection of azoxymethane (AOM, 10 mg/kg) and they began to receive 2.5% dextran sulfate sodium (DSS) in drinking water for 3 consecutive days followed by drinking water for 1 day. Mice were then submitted to three additional cycles of DSS (4 days with 2.5% DSS for the first one and 5 days with 1.5% DSS for the two others) with a resting period of 14 days between each cycle. Animals were sacrificed 85 days after AOM injection (day 124). At this time, each group contained the following number of animals: ND (n = 5), ND + AOM/DSS (n = 6), HCD (n = 7) and HCD + AOM/DSS (n = 9). At day 124, tumor burden was

monitored via endoscopy (**B**) and the total tumors numbers (**C**), the number of tumors per grade (**D**) and the percentage of mice ranked by the most severe grade (**E**) were then determined. * $p < 0.05$: student's *t*-test. (**F**) Relative expression of c-myc, Cyclin D1, Cox-2 and IkBα were assessed by RT-qPCR. Values are normalized to RPLP0. ns: non-significant, ** $p < 0.01$, *** $p < 0.001$, **** $p < 0.0001$: Bonferroni's multiple comparisons test. (**G–H**) Colon length was determined just after surgery. ns: non-significant, * $p < 0.05$, ** $p < 0.01$: Bonferroni's multiple comparisons test.

2.2. UNC5A, UNC5B and UNC5C Expression Is Downregulated in Colon Tumors Only in Mice Subjecting to the High Carbohydrate Diet

We then examined the colonic expression of UNC5A, UNC5B, UNC5C and UNC5D by RT-qPCR analysis (Figure 2A). Surprisingly and in disagreement with their tumor suppressor gene status, we did not observe any difference in the expression of UNC5A, UNC5B, and UNC5C between control group (ND) and AOM/DSS group under Normal Diet (ND + AOM/DSS) and UNC5D was found to increase (Figure 2A). These results suggest that downregulation of these family of genes is not necessary for CRC emergence. In contrast, our results showed a marked decrease in the level of UNC5A, UNC5B and UNC5C transcripts in the colon of AOM/DSS treated mice fed HCD (Figure 2A, compare ND or ND + AOM/DSS vs. HCD + AOM/DSS) strongly suggesting that nutrition is involved in the epigenetic downregulation of these genes that might occur during the progression of CRC.

2.3. O-GlcNAcylation Levels Are Enhanced in Colon from Mice Fed HCD and in Response to AOM/DSS Treatment

To test whether *O*-GlcNAcylation could be one of the molecular relays between this nutrition-dependent downregulation of UNC5A, UNC5B and UNC5C in AOM/DSS treated mice fed HCD, we performed immunohistochemistry experiments (Figure 2B,C). A 2.5 fold increase in IHC (Immunohistochemistry) score in tumors compared to normal tissues were observed in mice fed ND thus demonstrating that enhanced *O*-GlcNAcylation levels seem to be also a hallmark of CRC in mice (Figure 2B,C, compare ND vs. ND + AOM/DSS). In non-tumorigenic mice colons, the results showed that the HCD caused a 2 fold increase in colonic *O*-GlcNAcylation levels compared to the normal diet (Figure 2B,C, compare ND vs. HCD) thus confirming our previous work [19]. In addition, *O*-GlcNAcylation levels are further increased when the HCD is combined to the AOM/DSS treatment (Figure 2B,C, compare HCD vs. HCD + AOM/DSS). Nevertheless, the *O*-GlcNAcylation levels are not statistically higher in tumorigenic mice fed HCD compared to ND. This last observation argues against a direct link between *O*-GlcNAcylation levels and the downregulation of the members of the UNC5 family observed in tumorigenic mice fed HCD. However, and in an inverse correlation with the *O*-GlcNAcylation levels, our results show that subjecting mice to the High Carbohydrate Diet is sufficient to decrease the expression of UNC5A but not of the other members of the *UNC5* family in non-tumorigenic mice colons (Figure 2A, compare ND vs. HCD) and that this decrease in UNC5A transcripts is even greater in mice treated with AOM/DSS and subjected to the HCD (Figure 2A, compare HCD vs. HCD + AOM/DSS). Therefore, we hypothesized that *O*-GlcNAcylation could be one of the molecular relays between High Carbohydrate Diet and downregulation of the colonic expression of UNC5A exclusively.

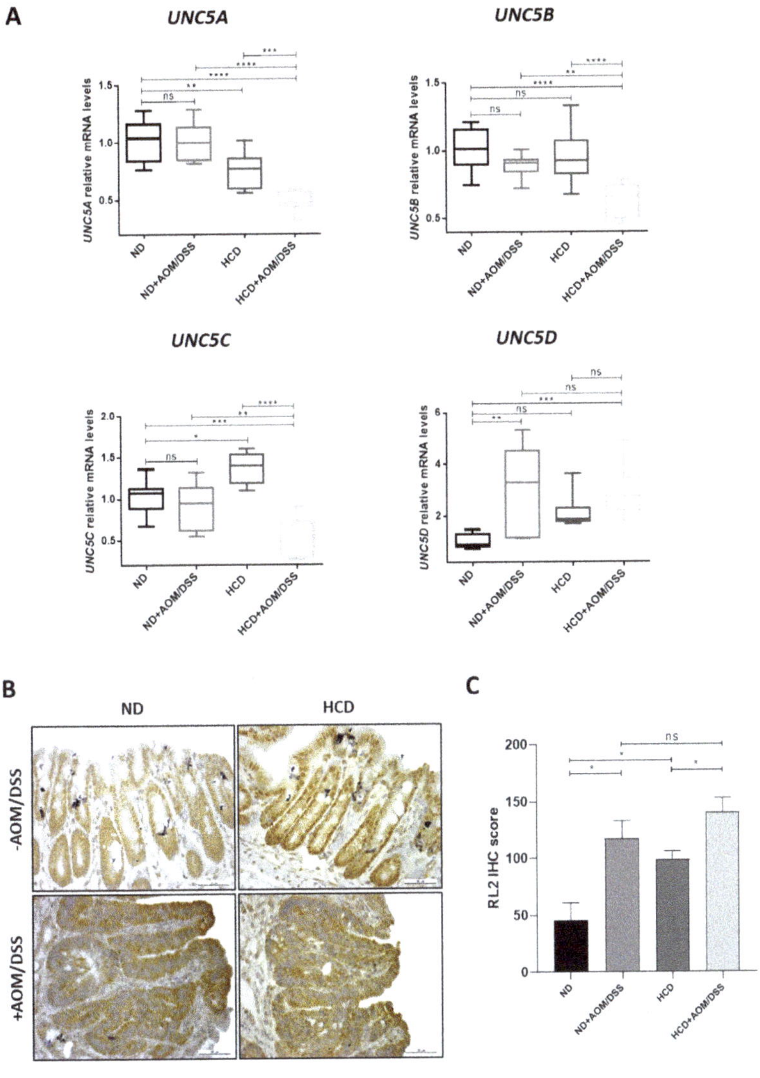

Figure 2. *UNC5A*, *UNC5B* and *UNC5C* expression is downregulated in colon tumors only in mice subjected to a High Carbohydrate Diet in association with an increase in O-GlcNAcylation levels. (**A**) Relative expression of UNC5A, UNC5B, UNC5C and UNC5D were assessed by RT-qPCR in the colon of mice fed a normal diet (ND) or a High Carbohydrate Diet (HCD) in combination or not with AOM/DSS. Values are normalized to RPLP0. ns: non-significant; * $p < 0.05$, ** $p < 0.01$, *** $p < 0.001$, **** $p < 0.0001$: Bonferroni's multiple comparisons test. (**B**) Tissue samples processed for immunohistochemistry staining of RL2. Pictures shown are representative of two sections per mouse. (**C**) IHC score performed in (**B**) was scored as described in the methods section and plotted as mean ± sem. * $p < 0.05$: unpaired *t*-test.

2.4. O-GlcNAcylation Is Involved in the Regulation of UNC5A Gene Expression in Human Colon Cancer HCT116 Cells

To evaluate the role of *O*-GlcNAcylation in the epigenetic downregulation of the *UNC5* family, we first examined the expression of UNC5A, UNC5B, UNC5C and UNC5D transcripts in the human colon cancer cell line HCT116 (Figure 3A). We detected significant levels of only UNC5A and UNC5B transcripts. We then investigated the effect of the knock-down of OGT by siRNA on the levels of the *UNC5* family members in this cell line. siOGT efficiency was checked at protein (Figure 3B) and mRNA level (Figure 3C) and we found that it induced a significant increase in UNC5A mRNA levels but had no effect on UNC5B (Figure 3C). Moreover, UNC5C and UNC5D transcripts remained undetectable upon OGT silencing (Decourcelle, A.; Dehennaut, V.; Université de Lille, CNRS, Inserm, CHU Lille, UMR9020-U1277—CANTHER—Cancer Heterogeneity, Plasticity and Resistance to Therapies, F-59000 Lille, France. Personal observation, 2020.). This first observation supports our hypothesis that *O*-GlcNAcylation is only involved in the regulation of the expression of UNC5A but not of the other members of the family. mRNA amounts can be regulated either at the transcriptional level or at the post-transcriptional level (mRNA stability). In order to investigate whether siOGT-induced increase of *UNC5A* expression was due to transcriptional regulation, we used a previously described *UNC5A* promoter activity reporter plasmid [36,37] in luciferase assay. As shown in Figure 3D, siOGT caused a 2-fold induction of the *UNC5A* promoter activity correlated with a decrease in *O*-GlcNAcylation levels (Figure 3B). Inversely, treating HCT116 cells with the potent OGA inhibitor Thiamet G (TG) induced an *O*-GlcNAcylation increase (Figure 3E) correlated with a drop of 25% in UNC5A mRNA levels (Figure 3F) and promoter activity (Figure 3G). In contrast, Thiamet G treatment has no effect on UNC5B mRNA levels (Figure 3F). Taken together, these data demonstrate that *O*-GlcNAcylation is involved in the regulation of the transcription of colonic *UNC5A*.

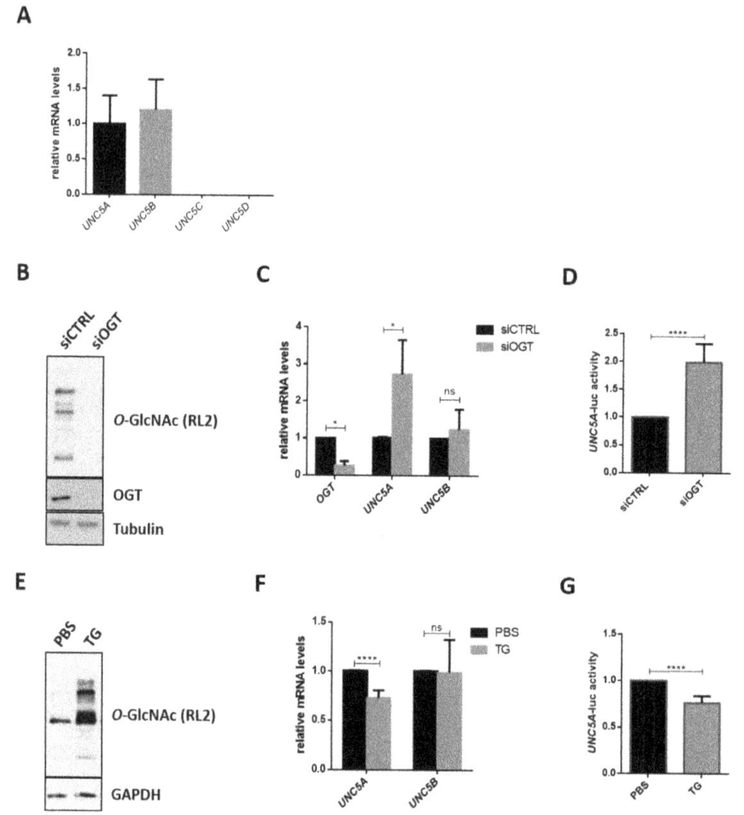

Figure 3. *O*-GlcNAcylation is involved in the regulation of the transcription of *UNC5A* in human colon cancer HCT116 cells. (**A**) Expression of the four members of the *UNC5* gene family (*UNC5A*, *UNC5B*,

UNC5C and *UNC5D*) was analyzed by RT-qPCR in HCT116 cells. Data shown are the average ± SD of three independent experiments. (**B–D**) HCT116 cells were transfected with non-target control siRNA (siCTRL) or with OGT siRNA (siOGT) for 72 h. (**B**) To ensure siRNA efficiency, a fraction of the cell lysates was analyzed by Western blot (WB) with anti-*O*-GlcNAc (RL2) and anti-OGT antibodies. Tubulin was used as a loading control. Data shown are representative of three independent experiments. (**C**) Cells were harvested for total RNA extraction. The mRNA expression levels of OGT, UNC5A and UNC5B were assessed by RT-qPCR. Values were normalized to *RPLP0*. Data shown are the average ± SD of three independent experiments. ns: non-significant, * $p < 0.05$: unpaired *t*-test. (**D**) Then, 24 h after siRNA transfection, cells were transfected with 200 ng of an *UNC5A* promoter activity reporter plasmid and 25 ng of β-galactosidase reporter. Thus, 48 h later, *UNC5A*-luciferase activity was determined as described in the experimental procedures section. Data shown are the average ± SD of three independent experiments. **** $p < 0.0001$: unpaired *t*-test. (**E–G**) HCT116 cells were treated with the OGA inhibitor Thiamet G (TG) (1 μM final in PBS (Phosphate Buffered Saline)) for 24 h. (**E**) Treatment efficiency was ensured by WB analyses of *O*-GlcNAcylation (RL2) levels. GAPDH (Glyceraldehyde-3-phosphate dehydrogenase) was used as a loading control. Data shown are representative of three independent experiments. (**F**) mRNA expression level of UNC5A and UNC5B were assessed by RT-qPCR. Values were normalized to *RPLP0*. Data shown are the average ± SD of three independent experiments. **** $p < 0.0001$: unpaired *t*-test. (**G**) Cells were transfected with 200 ng of an *UNC5A* promoter activity reporter plasmid and 25 ng of β-galactosidase reporter. Following this, 24 h later, cells were treated with TG (1 μM final in PBS) for 24 h. *UNC5A*-luciferase activity was determined as described in the experimental procedures section. Data shown are the average ± SD of three independent experiments. **** $p < 0.0001$: unpaired *t*-test.

2.5. *O-GlcNAcylation Regulates the PRC2-Mediated Repression of UNC5A in Human Colon Cancer Cells*

As mentioned in the introduction section, several members of the Polycomb Repressive Complex 2 (PRC2), including the histone methyl transferase EZH2, are overexpressed in CRC [26–28]. Moreover, several studies have demonstrated that OGT and *O*-GlcNAcylation regulate EZH2 functions [29–33]. Next, we wondered whether the PRC2 complex was also involved in the regulation of *UNC5A* transcription in colon cancer cells. To answer this question, we first inhibited the methyl transferase activity of EZH2 in HCT116 cells with GSK343, a selective SAM (S-adenosyl methionine)-competitive EZH2 inhibitor (Figure 4A) and analyzed the levels of UNC5A mRNA (Figure 4B) and promoter activity (Figure 4C). GSK343 clearly diminished the level of H3K27 trimethylation in cells (Figure 4A) and increased *UNC5A* expression (Figure 4B) and promoter activity (Figure 4C). These first results thus suggest that EZH2 is also involved in the regulation of *UNC5A* expression in HCT116 cells. To confirm this result, we performed siRNA knockdown of EZH2 alone or in combination with siOGT in HCT116 cells and analyzed the levels of UNC5A mRNA (Figure 4D). We first observed that siEZH2 induced a derepression of *UNC5A* expression in line with the results obtained with GSK343 inhibitor. Moreover, siEZH2 increased UNC5A transcripts at a level similar to siOGT and the simultaneous knockdown of OGT and EZH2 had no additive effect compared to siEZH2 or siOGT alone. It is also to note that the abolition of the expression of EZH2 or the inhibition of its catalytic activity had no effect on OGT expression nor on *O*-GlcNAcylation levels (Figure 4A,D,E) arguing against an indirect modulation of *UNC5A* expression through the modulation of OGT functions in these conditions. This last result thus suggests that OGT and EZH2 act together to repress *UNC5A* transcription. We confirmed this result by inhibiting OGT and/or EZH2: respectively with Ac5S-GlcNAc (Ac5S) and GSK343 specific inhibitors (Figure 4E) and validated them also in LS174T human colorectal cancer cell line (Supplementary Figure S2). We then overexpressed the catalytic core of PRC2 complex (composed of EZH2, SUZ12 and EED) in HCT116 cells by co-transfecting cells with plasmids encoding Myc-EZH2, HA-SUZ12 and HA-EED (Figure 5A, EES). In PRC2-transfected cells, we observed a 40% reduction of the *UNC5A* promoter activity compared to mock-transfected ones (Figure 5A, right panel) thus confirming that *UNC5A* is a target of the PRC2 complex. However, this repression was no longer observed in cells co-transfected with siOGT suggesting that *O*-GlcNAcylation is required for the PRC2-mediated repression of *UNC5A*. In a second set of experiments, we again overexpressed the core PRC2, this time in conjunction with treatment of the HCT116 cells with Ac5S-GlcNAc. We then analyzed the UNC5A mRNA levels by RT-qPCR (Figure 5B). In agreement with the luciferase activity assay, overexpression of the core PRC2 led to a decrease in UNC5A mRNA and this was prevented by simultaneous treatment of cells with Ac5S-GlcNAc (Figure 5B, right panel). Taken together, these results demonstrate that *O*-GlcNAcylation regulates the PRC2-mediated repression of *UNC5A* in human colon cancer cells.

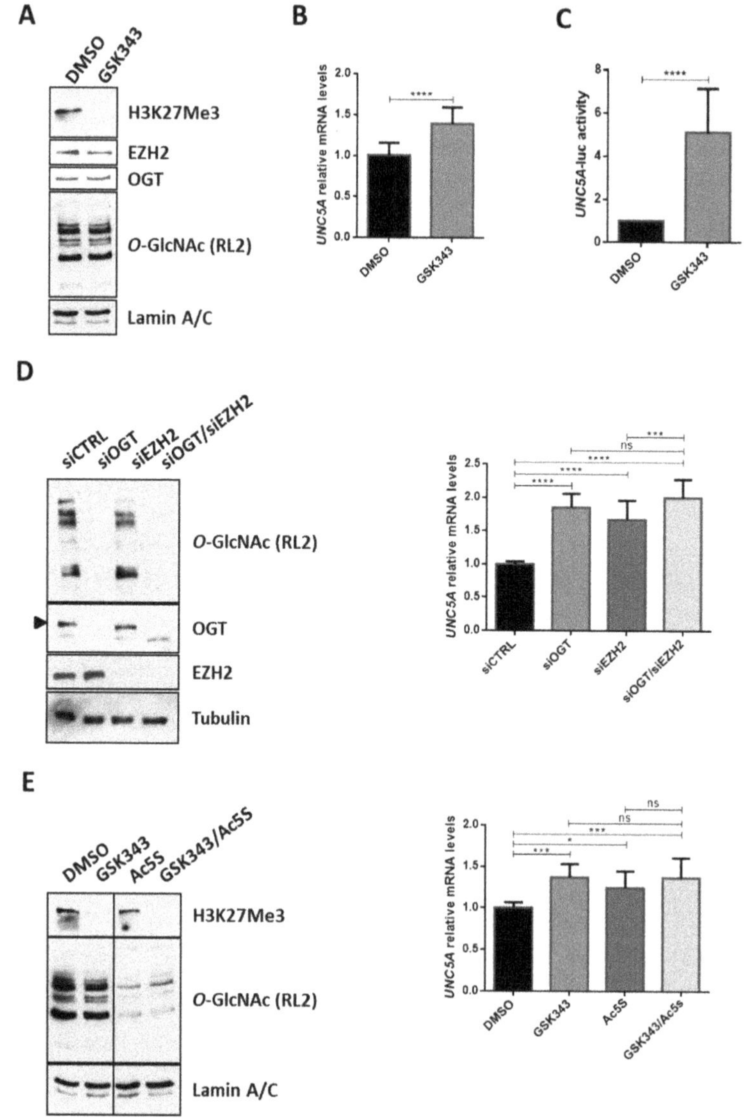

Figure 4. EZH2 and OGT act together to repress the transcription of *UNC5A* in colon cancer cells. (**A**,**B**) HCT116 cells were treated with the EZH2 inhibitor GSK343 (5 μM final in DMSO (Dimethyl sulfoxide)) for 48 h. (**A**) Treatment efficiency was ensured by WB analyses of H3K27Me3 levels from

chromatin-bound proteins extracts. EZH2, OGT and *O*-GlcNAcylation (RL2) levels were also assessed. Lamin A/C were used as a loading control. Data shown are representative of three independent experiments. (**B**) mRNA expression level of UNC5A was assessed by RT-qPCR. Values were normalized to RPLP0. Data shown are the average ± SD of three independent experiments. **** $p < 0.0001$: unpaired *t*-test. (**C**) Cells were transfected with 200 ng of an *UNC5A* promoter activity reporter plasmid and 25 ng of β-galactosidase reporter. Thus, 6 h later, cells were treated with GSK343 (5 μM final in DMSO) for 48 h. *UNC5A*-luciferase activity was determined as described in the experimental procedures section. Data shown are the average ± SD of three independent experiments. **** $p < 0.0001$: unpaired *t*-test. (**D**) HCTT116 cells were transfected either with non-target siRNA control (siCTRL), OGT siRNA (siOGT), EZH2 siRNA (siEZH2) or with a combination of siOGT and siEZH2 for 72 h. Cells were harvested either for total proteins or RNA extraction. Left panel: WB analyses were performed using the indicated antibodies. Data shown are representative of five independent experiments. Right panel: mRNA expression level of *UNC5A* was assessed by RT-qPCR. Values were normalized to *RPLP0*. Data shown are the average ± SD of five independent experiments. ns: non-significant, *** $p < 0.001$, **** $p < 0.0001$: Bonferroni's multiple comparisons test. (**E**) HCTT116 cells were treated with the EZH2 inhibitor GSK343 (5 μM final in DMSO) or the OGT inhibitor Ac5S-GlcNAc (50 μM final in DMSO) alone or in combination. Cells were harvested either for total protein or RNA extraction. Left panel: WB analyses of Lamin A/C, H3K27Me3 and *O*-GlcNAcylation levels from chromatin-bound proteins extracts. Data shown are representative of four independent experiments. Right panel: amount of *UNC5A* transcripts was assessed by RT-qPCR. Values were normalized to *RPLP0*. Data shown are the average ± SD of four independent experiments. ns: non-significant, * $p < 0.05$, *** $p < 0.001$: Bonferroni's multiple comparisons test.

2.6. *O-GlcNAcylation Drives the Recruitment of EZH2 onto the UNC5A Promoter*

We then investigated the mechanisms by which *O*-GlcNAcylation could influence the EZH2 mediated-repression of *UNC5A*. Indeed, regulation of EZH2 functions by its *O*-GlcNAcylation has been evidenced in several studies conducted in different cell lines but the exact roles of this glycosylation are not so clear. Indeed, while some studies argue for a role of *O*-GlcNAcylation in the regulation of EZH2 stability and catalytic activity [29–31], others propose that the glycosylation rather regulates EZH2 recruitment to some of its target genes such as *FOXC1* in breast cancer cells [32] or *IL-15* in muscle [33]. First, we checked the modification of endogenous EZH2 by *O*-GlcNAcylation in HCT116 cells by sWGA (succinyl Wheat Germ Agglutinin)-beads enrichment experiments (Figure 6A). Briefly, HCT116 cells were treated with Thiamet G for 24 h to maintain high *O*-GlcNAcylation levels in the cells (Figure 6A, bottom panel). Total cell lysates were incubated with sWGA-beads to extract *O*-GlcNAcylated proteins or, as a negative control, with sWGA-beads preincubated with free GlcNAc (Figure 6A, top panel). EZH2 was easily detected in the sWGA-enriched fractions (with or without TG) but not in the negative controls thus proving *O*-GlcNAcylation of EZH2 in HCT116 cells. Inhibition of OGT activity by Ac5S-GlcNAc did not affect EZH2 expression, whereas DZNEP, a well-known EZH2 destabilizing agent, did (Figure 6B). Similarly, the suppression of OGT expression by siRNA did not lead, as might be expected, to a decrease in EZH2 protein levels (Figures 4D and 6C,D Inputs). In addition, increasing *O*-GlcNAcylation levels with Thiamet G did not lead to any variation in EZH2 expression (Figure 6A,E Inputs). All these results therefore do not argue for a stabilizing role of *O*-GlcNAcylation in our study model. *O*-GlcNAcylation also does not seem to influence the methyltransferase activity of EZH2 in this cell line since the inhibition of OGT activity or expression had no impact on the levels of H3K27Me3, the epigenetic mark deposited by the PRC2 complex (Figures 4E and 6C). Among the EZH2 *O*-GlcNAcylation sites identified to date, Ser^{313} is located in the domain of interaction with SUZ12 [38]. Nevertheless, co-immunoprecipitation experiments between EZH2 and SUZ12 carried out in HCT116 cells under *O*-GlcNAcylation levels modulation conditions showed that this post-translational modification did not seem to influence the interaction between these two partners (Figure 6D,E). We finally analyzed whether *O*-GlcNAcylation could drive the recruitment of the PRC2 complex onto the *UNC5A* promoter. To test this hypothesis, we performed CUT&RUN experiments (an alternative method to ChIP that also enables mapping of protein-DNA interactions) under conditions of modulation of *O*-GlcNAcylation levels (Figure 6F,G). In this way, we showed that EZH2 was bound onto the *UNC5A* promoter in HCT116 cells and that inhibition of OGT with Ac5S-GlcNAc clearly diminished its binding (Figure 6F). On the contrary, OGT inhibition led to an increased binding of the H3K4Me3 activating mark on the *UNC5A* promoter (Figure 6G). So, taken together, our overall results demonstrate that in colon cancer cells, *O*-GlcNAcylation of EZH2 targets the PRC2 complex onto the *UNC5A* promoter to inhibit its transcription.

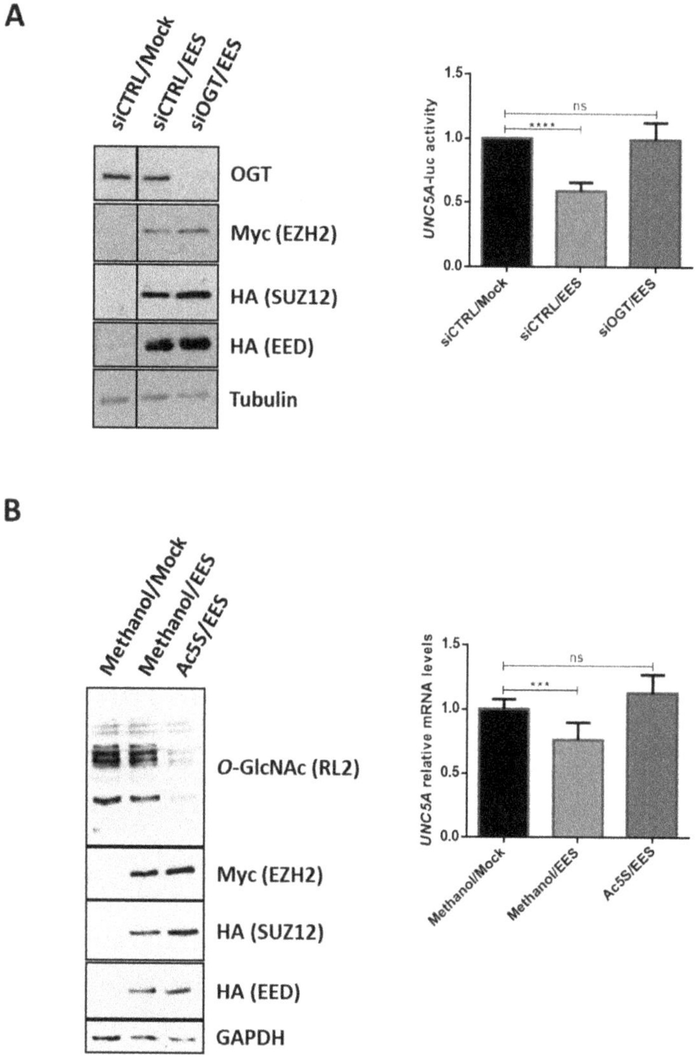

Figure 5. *O*-GlcNAcylation mediates the PRC2-mediated repression of *UNC5A*. (**A**) HCT116 cells were co-transfected with the core PRC2 complex (EES: Myc-EZH2, HA-EED, HA-SUZ12), 200 ng of plasmid encoding the luciferase reporter gene under the control of the *UNC5A* promoter and 25 ng of β-galactosidase in combination or not with siOGT. Left panel: transfection efficiency and OGT inhibition

were checked by WB analyses with the indicated antibodies. Data shown are representative of three independent experiments. Right panel: luciferase activity was determined 48 h later and was normalized to β-galactosidase activity. Data shown are the average ± SD of three independent experiments. ns: non-significant, **** $p < 0.0001$: Bonferroni's multiple comparison tests. (**B**) Cells were co-transfected with plasmids encoding the core PRC2 (EES: Myc-EZH2, HA-EED, HA-SUZ12), or with empty vectors (Mock). Thus, 24 h after transfection, cells were treated with the potent OGT inhibitor Ac5s-GlcNAc (50 μM final in methanol) for 24 h. Left panel: transfection efficiency and OGT inhibition were checked by WB analyses with the indicated antibodies. Data shown are representative of three independent experiments. Right panel: mRNA expression level of UNC5A was assessed by quantitative RT-PCR. Values were normalized to RPLP0. Data shown are the average ± SD of three independent experiments. ns: non-significant, *** $p < 0.001$: Bonferroni's multiple comparison tests.

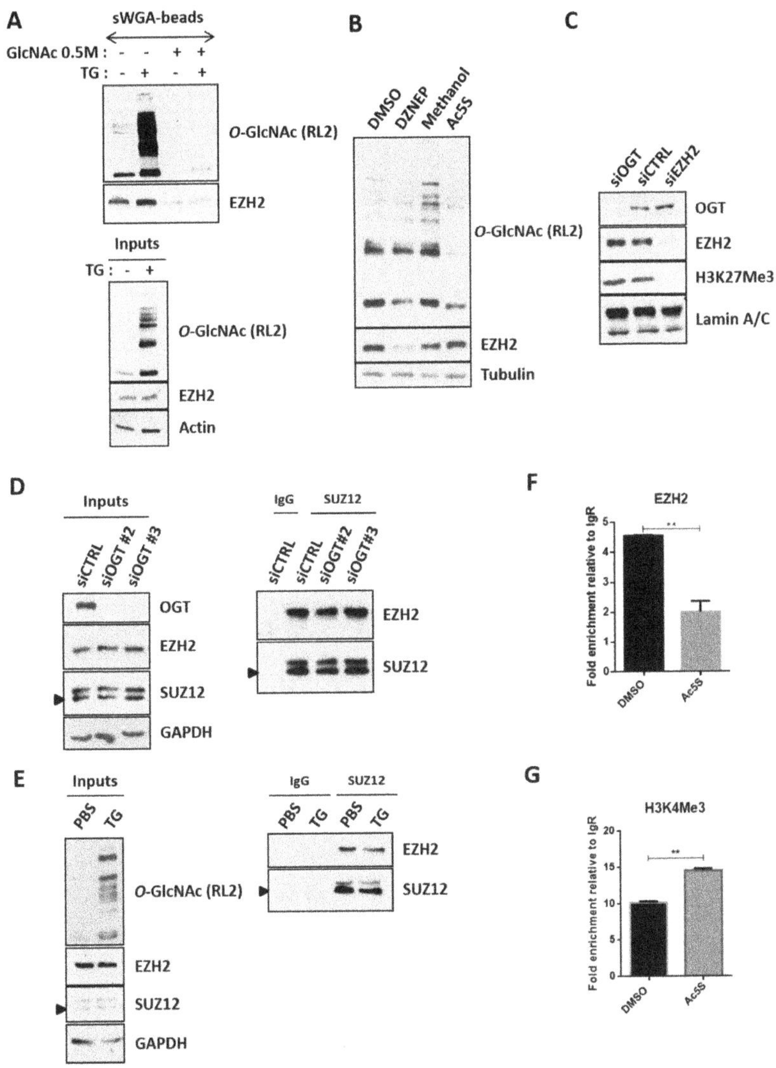

Figure 6. In colon cancer cells, *O*-GlcNAcylation neither influences EZH2 stability, methyltransferase activity nor its interaction with SUZ12 but drives the PRC2 recruitment onto the *UNC5A* promoter.

(**A**) Enrichment of the *O*-GlcNAcylated proteins from Thiamet G-treated (1 µM final in PBS, 24 h) or non-treated cells were performed with sWGA-agarose beads as described in the experimental procedures section. In total, 25µg of whole cell lysates (Inputs) and bound proteins were analyzed by WB with anti-EZH2 and anti-*O*-GlcNAc (RL2) antibodies. Actin was used as a loading control. Data shown are representative of two independent experiments. (**B**) HCT116 cells were treated with the EZH2 destabilizing agent DZNEP (3-Deazaneplanocin A) (5 µM final in DMSO) or the OGT inhibitor Ac5S-GlcNAc (50 µM final in methanol) respectively for 48 h or 24 h. WB analyses were performed using the indicated antibodies. Tubulin was used as a loading control. Results are representative of at least three independent experiments. (**C**) HCTT116 cells were transfected either with non-target siRNA control (siCTRL) or siRNA targeting OGT (siOGT) or EZH2 (siEZH2, used as a positive control) for 72 h. Chromatin-bound proteins were extracted and WB analyses were performed using the indicated antibodies. Lamin A/C were used as a loading control. Results are representative of at least three independent experiments. (**D–E**) HCT116 cells were transfected with 2 different OGT-targeting siRNA (siOGT#2 or siOGT#3) or a non-relevant siRNA (siCTRL) for 72 h (**D**) or were treated with 1 µM Thiamet G for 24 h (**E**). The interaction between SUZ12 and EZH2 was then evaluated by co-immunoprecipitation experiments. Results are representative of three independent experiments. (**F–G**) HCT116 cells were treated with the OGT inhibitor Ac5S-GlcNAc (50 µM final in DMSO) for 24 h. Then binding of EZH2 (**F**) as well as presence of the H3K4Me3 activating mark (**G**) onto the *UNC5A* promoter were assessed by CUT&RUN experiments. Data shown represent the fold enrichment relative to the isotype control (Rabbit IgG) of a single experiment deposited in duplicate and is representative of two independent experiments. ** $p < 0.01$: unpaired *t*-test.

3. Discussion

In recent decades, changes in Western lifestyle (increased sedentary not compensated by a decrease in caloric intake) have largely contributed to the increased incidence of CRC. In this regard, several epidemiological studies have shown that there is a higher risk factor for developing CRC in patients with metabolic syndrome, type-2 diabetes or obesity [16–18]. In addition to the well-known genetic origin of CRC, many studies have shown that colon carcinogenesis also involves alterations in the epigenetic regulation of genome. Furthermore, a growing number of studies tends to prove that the epigenome is able to integrate nutritional informations and it seems therefore obvious that nutritional intake, by modifying the epigenome, could influence the emergence and progression of CRC [1–3]. However, the underlying mechanisms are still insufficiently known. The nucleotide sugar UDP-GlcNAc, donor for the *O*-GlcNAc modification, is synthetized through the Hexosamine Biosynthesis Pathway (HBP) at the crossroad of glucose, amino acid, fatty acid and nucleotide metabolisms. UDP-GlcNAc and *O*-GlcNAcylation are therefore considered as sensors of the nutritional state of the organism [15] which can relay the effects of an excessive food supply, unbalanced diet, obesity and other metabolic problems. It has been previously shown that the expression of the core PRC2 complex (EZH2, EED and SUZ12) that catalyzes the deposit of the epigenetic repressive mark H3K27Me3 [26–28] as well as OGT and *O*-GlcNAcylation levels are increased during colorectal carcinogenesis [19–22]. Inversely, the expression of the tumor suppressor genes of the *UNC5* family is frequently downregulated in CRC in part through epigenetic mechanisms not fully deciphered [6–10]. In this study, in agreement with the fact that poor eating habits may influence colon carcinogenesis, we observed that consumption of a High Carbohydrate Diet (HCD) worsens colon carcinogenesis in mice. In this model, we found that the expression of UNC5A, UNC5B and UNC5C was diminished in tumors of mice only fed HCD but not of mice fed ND thus linking nutrition to their repression during the progression of CRC. We then tested the hypothesis that O-GlcNAcylation could be one of the molecular relays between this nutrition-dependent repression of UNC5A, UNC5B and UNC5C. Our in vivo *and* in vitro results argued against a direct correlation between *O*-GlcNAcylation levels and expression of UNC5B and UNC5C but strongly suggested a more direct link between *O*-GlcNAcylation and UNC5A expression. Mechanistically, we provide evidences that the *O*-GlcNAcylated form of EZH2 prevents the transcription of *UNC5A* in human colon cancer cells through aberrant *O*-GlcNAcylation and abnormal targeting of the PRC2 complex onto its promoter, thus linking nutrition to downregulation of *UNC5A* in CRC. However, further work will be required to understand the molecular link between nutrition and epigenetic regulation of the other members of the UNC5 family.

O-GlcNAcylation of EZH2 had already been described in several cell lines but the functions of this post-translational modification are still uncleared. In fact, some studies conclude that *O*-GlcNAcylation regulates EZH2 stability and methyltransferase activity [29–31,38] while others tempt to demonstrate that it rather steers the PRC2 complex onto specific loci [32,33]. Here, we confirmed the previously described *O*-GlcNAcylation of EZH2 in the colon cancer cell line HCT116. In our hands, *O*-GlcNAcylation does not seem to affect either stability or catalytic activity of EZH2 but appears to regulate the binding of EZH2 at specific loci including the *UNC5A* promoter in colon cancer cells.

UNC5 receptors belong to the family of the so-called "dependence receptors". Such receptors initiate two opposite signaling pathways. When ligand (Netrin-1 for UNC5) is available, these receptors transduce a "positive signal" leading to cellular proliferation, differentiation, migration or survival. In absence of their ligand, they are still active but rather induce a "negative signal" that triggers caspase-dependent apoptosis [5]. Several studies have demonstrated that constitutive inhibition of the death signal induced by these receptors contributes to cell transformation. Conversely, their re-expression could represent a protective mechanism that limits tumor development through apoptosis induction of tumor cells, thus defining the members of the *UNC5* family as tumor suppressor genes. For example, in mice, inactivation of *UNC5C* is associated with increased intestinal tumor progression and decreased tumor cell apoptosis [8]. In human cancer cells, enforced expression of UNC5A, UNC5B and UNC5C inhibits cell-anchorage growth and invasion in a way related to their pro-apoptotic activity [7]. *UNC5A*, *UNC5B* and *UNC5D* are transcriptionally regulated by the tumor suppressor p53 and are involved in the p53-dependent apoptosis in response to DNA damages induced by conventional chemotherapeutic drugs like doxorubicin [39–42]. However, in an elegant paper, Paradisi et al. demonstrated, in a variety of cancer cell lines, that doxorubicin, 5-fluorouracil, paclitaxel and cisplatin treatments induced not only an increase of UNC5 receptors but also a concomitant p53-dependent increase of Netrin-1 thus preventing the pro-apoptotic action of UNC5 receptors [43]. In the same study, the authors also showed that interfering with Netrin-1 expression by siRNA or using two Netrin-1/Netrin-1 receptor inhibitors of interaction (TRAP-netrin^{DCC}, TRAP-netrin^{UNC5A}) potentiates doxorubicin induced cell death. According to these results, they proposed Netrin-1 upregulation as a survival mechanism of cancer cells in response to these drugs. Netrin-1 interference combined to conventional therapies has been then envisaged as a promising therapeutic approach for tumors resistant to chemotherapy. Interestingly, we observed that the siRNA knockdown of OGT led to increased *UNC5A* expression but had no effect on *Netrin-1* mRNA levels (data not shown). So, in the near future, it would be of great interest to go further into the mechanisms of transcriptional regulation of *UNC5A* by the *O*-GlcNAcylated form of EZH2 in response to chemotherapy and one could propose targeting with OGT/EZH2 interaction to enhance efficiency of chemotherapy in CRC.

4. Materials and Methods

4.1. AOM/DSS-Induced CRC Model

All the procedures were carried out according to the French guidelines for the care of experimental animals and the experimental procedure was approved by the Animal Care Committee of the French Research Ministry (Autor. APAFiS #1879-2018121918307521). Male C57BL/6JRj mice (n = 28; 8 weeks old; Charles River Saint-Germain sur l'Arbresle) were maintained under controlled room temperature, humidity and light (12/12 h light/dark cycle) with free access to food and tap water. Half of the mice were subjected to a Normal Diet (ND; standard laboratory chow; 16.1% protein, 3.1% lipid, 60.4% NFE (Nitrogen-Free Extract), 4.6% minerals and 3.9% fibers; A04; SAFE) and the other half to a High Carbohydrate Diet (HCD; 13.5% protein, 3.2% lipid, 76.8% NFE, 3.6% minerals, 0.7% cellulose and 1.1%

starch; U8960P version 0002; SAFE) (Figure 1A). Thirty-nine days after the beginning of the two diets, we used the AOM/DSS method to induce inflammation-driven CRC [35]. Animals received a single intraperitoneal injection of azoxymethane (AOM, 10 mg/kg, A5486 Sigma) and they began to receive 2.5% dextran sulfate sodium (DSS, 0216011080 MP Biomedicals) in drinking water for 3 consecutive days followed by drinking water for 1 day. Mice were then submitted to three additional cycles of DSS (4 days with 2.5% DSS for the first one and 5 days with 1.5% DSS for the two others) with a resting period of 14 days between each cycle. Weight and blood glucose levels were monitored every 2 weeks during all the time course of the experiment. At the end of the experiment, tumor burden was monitored via endoscopy and colons were collected and washed in Phosphate Buffered Saline (PBS). Samples were kept at −80 °C for further RNA extraction. Part of the samples was also fixed in 10% buffered formalin overnight for further immunohistochemistry (IHC) analysis.

4.2. *Assessment of Tumorigenesis Using Colonoscopy*

Tumor developments were assessed using a high-resolution Karl Storz colonoscope (1.9 mm outer diameter; Tuttlingen, Germany) at the end of protocol (day 124). Mice were anesthetized using isoflurane for the duration of the procedure and closely monitored. For tumors assessment we followed the method described by Becker et al. [44]. Tumors observed from colonoscopy videos were counted, measured and categorized into size: grade 1 (very small but detectable tumor), grade 2 (tumor covering up to one eighth of the colonic circumference), grade 3 (tumor covering up to a quarter of the colonic circumference), grade 4 (tumor covering up to half of the colonic circumference) and grade 5 (tumor covering more than half of the colonic circumference).

4.3. *Immunohistochemistry Staining and Quantification*

After fixation, tissues were embedded in paraffin wax by automatic sample preparation system (LOGOS One, Milestone). Serial histological sections of 4 μm thickness were cut, deparaffinized and rehydrated. For antigen unmasking, sections were placed in 10 mM sodium citrate buffer pH 6.0 and incubated in a heat induced antigen retrieval chamber for 20 min at 121 °C. After washing, sections were blocked for 30 min with 5% BSA (Bovine Serum Albumin) in PBS. The following primary antibody was then incubated overnight at 4 °C: anti-*O*-GlcNAc (RL2) (NB300-524, Novus Biologicals) used at a concentration 1/1000. After washing, tissue sections were incubated 1 h at room temperature with the specified secondary antibody bound to HRP (Horseradish Peroxidase): Anti-Mouse IgG—Horseradish PeroxidaseLinked Species—Specific Whole Antibody (from sheep) (#NA931, GE Healthcare). Samples were washed with PBS and the signal was subsequently revealed with DAB (Diaminobenzidine) (#8059, Cell Signaling). Finally, tissue sections were counterstained with hematoxylin. Images were acquired with a DM5500B microscope (Leica Microsystems, Nanterre, France) and mucosal layers were photographed at a magnification of x20. The assessment of staining was performed using ImageJ software and the IHC score = staining intensity score (low:1, medium:2 and high:3) × % of surface staining.

4.4. *Cell Culture*

HCT116 cells purchased from ATCC were maintained in Mc Coy's 5A (modified) medium with Glutamax (Thermofischer Scientific) supplemented with 10% fetal calf serum and 1% ZellShield™ (Biovalley). Cells were cultured at 37 °C in water-saturated 5% CO_2 atmosphere.

4.5. *Inhibitors*

The EZH2 inhibitor, GSK343 (Sigma-Aldrich), was dissolved in dimethyl sulfoxide (DMSO) at 10 mM as stock solution and used at a final concentration of 5 μM. The EZH2 destabilizing agent, DZNEP (Sigma-Aldrich), was dissolved in DMSO at 10 mM as stock solution and used at a final concentration of 5 μM. The OGA inhibitor, Thiamet G (TG, Sigma-Aldrich), was dissolved in PBS at 1 mM as stock solution and used at a final concentration of 1 μM. The OGT inhibitor, Ac5S-GlcNAc

(Ac5S, kind gift of GW. Hart and D. Vocadlo), was dissolved either in methanol or DMSO depending of their provenance at 50 mM as stock solution and used at a final concentration of 50 µM. Inhibitors were added to the cell culture medium for 24 h or 48 h.

4.6. *Plasmids and Transfection*

The expression vectors for Myc-EZH2, HA-EED and HA-SUZ12 have been previously described (24). Cells were seeded in 6 wells plates and transfected with 330 ng of each plasmid in 5 mL of complete culture medium by the Lipofectamine (Lipo2000, Invitrogen) method (4µL). Cells were harvested 48 h after transfection.

4.7. *Small Interfering RNA*

HCT116 cells were reverse-transfected with Lipofectamine RNAiMax (Invitrogen) according to manufacturer's instructions using 5 nM small interfering RNA targeting OGT (siGENOME human OGT siRNA D-019111-01, Dharmacon), EZH2 (EZH2 siGENOME SMART Pool M-004218-03-0005, Dharmacon) or a scrambled control sequence (siCTRL; siGENOME RISC free control siRNA, Dharmacon) as previously described [19]. Seventy-two hours later, cells were harvested for RNA/protein extraction.

4.8. *Quantitative RT-PCR*

RNA was isolated using Nucleospin® RNA mini spin kit (Macherey-Nagel) according to the manufacturer's instructions. 1µg of total RNA was reverse transcribed using random primers and MultiScribeTM reverse transcriptase (Applied Biosystems). Real-time PCR analysis was performed by Power SYBR Green (Applied Biosystems) in a MX3005P fluorescence temperature cycler (Stratagene) according to the manufacturer's instructions. Results were normalized with respect to *RPLP0* mRNA used as internal control. The primers used for the RT-qPCR analyses are summarized in Supplementary Table S1.

4.9. *Total and Chromatin-Bound Proteins Extraction, Western Blotting and Antibodies*

For total protein extraction, cells were lysed in RIPA (Radioimmunoprecipitation assay) buffer (10 mM Tris [pH 7.4], 150 mM NaCl, 1 mM EDTA, 1% Triton X-100, 0.5% sodium deoxycholate, 0.1% SDS and proteases inhibitors added at the time of preparation). For chromatin-bound proteins extraction, subcellular fractionation was performed as described in Füzesi-Levi et al. [45]. Protein concentration was determined using the Micro BCA (Bicinchoninic acid) Protein Assay Kit (Thermofisher Scientific). Equal amounts of proteins were separated by SDS-PAGE and transferred onto nitrocellulose membranes (GE Healthcare). After 1 h of blocking in PBSM (PBS with 5% milk), the membranes were incubated overnight at 4 °C with specific primary antibodies in PBSTM (PBSM with 0.1% Tween) and washed three times with PBSN (PBS with 0.1% NP-40). The membranes were next incubated for 1 h at room temperature with secondary antibodies coupled to peroxydase (Amersham) in PBSM, washed three times in PBSN and revealed by chemiluminescence.

Mouse monoclonal anti-*O*-GlcNAc (RL2) was purchased from Life technologies (MA1072). Rabbit polyclonal anti-OGT (DM17) was purchased from Sigma-Aldrich (#O6264). Mouse monoclonal anti-tubulin (sc-23948), mouse monoclonal anti-GAPDH (sc-32223), rabbit polyclonal anti-lamin A/C (sc-20681), rabbit polyclonal anti-actin (sc-1616-R) and rabbit polyclonal anti-Myc (sc-789) were purchased from Santa Cruz Biotechnology. Rabbit polyclonal anti-EZH2 (#5246) and rabbit polyclonal anti-H3K27Me3 (#9733) were purchased from Cell signaling. Rabbit polyclonal anti-SUZ12 was purchased from Abcam (ab12073). Mouse monoclonal anti-HA was purchased from Biolegend (MMS-101P). Anti-mouse IgG-HRP was purchased from GE Healthcare (NA931V). Donkey anti-rabbit IgG-HRP was purchased from Millipore (AP182P).

4.10. Luciferase Promoter Activity Assays

The pGl3-UNC5A promoter construct has been kindly provided by A. Paradisi (P. Mehlen team, Cancer research center of Lyon, Lyon, France). HCT116 cells, seeded in 12-wells plates, were transfected with 200 ng of the *UNC5A* promoter activity reporter plasmid and 25 ng of β-galactosidase reporter using Lipofectamine 2000 (Invitrogen) according to the manufacturer's instructions. Thus, 48 h after transfection, cells were lysed in Luciferase assay buffer (25 mM glycyl glycine [pH 7.8], 15 mM $MgSO_4$, 4 mM EGTA, 1% Triton X-100). Luciferase and β-galactosidase activities were measured by using, respectively, beetle luciferine (Promega) and the Galacto-light kit (Tropix) with a Berthold chemiluminometer. After normalization to the β-galactosidase activity, the data were expressed as the fold-change *UNC5A*-luciferase activity relative to control, which was given an arbitrary value of 1. A fraction of cell lysates was also subjecting to Western Blot (WB) analyses to ensure efficiency of cells treatments with siRNA or inhibitors.

4.11. Enrichment of O-GlcNAc-Bearing Proteins with sWGA Immobilized on Agarose BEADS

Cells were lysed in RIPA buffer and protein concentration was determined. For each condition, 2 mg of proteins were incubated either with 50 μL of succinylated-Wheat Germ Agglutinin (sWGA) agarose beads (Vector Laboratories) for 2 h at 4 °C or with sWGA beads beforehand incubated with 0.5 M free GlcNAc (to control the specificity of the reaction). sWGA-bound proteins were collected and washed three times with washing buffer (10 mM Tris [pH 7.4], 150 mM NaCl, 0.5 mM EDTA, 1% Triton X-100, 0.5% sodium deoxycholate, 0.2% SDS). Proteins were then eluted from the beads in 2X Laemmli buffer (50 mM Tris-HCl pH6.5, 2.5% SDS, 5% 2-mercaptoethanol, 10% glycerol and a hint of bromophenol blue) and resolved by SDS-PAGE.

4.12. Co-Immunoprecipitation

Twenty-four hours or 72 h respectively after Thiamet G treatment and siRNA transfection, cells were rinsed with cold PBS and lysed with 1 mL of cold IPH (Immunoprecipitation homogeneization) buffer (50 mM Tris pH8, 150 mM NaCl, 5 mM EDTA and 0.5% NP40) supplemented with proteases inhibitors. Cells were put on ice for 15 min and scraped before being placed in rotation at 4 °C for 15 min. Cell lysates were briefly sonicated (22%—2 s). After a centrifugation at 14,000 rpm for 15 min at 4 °C, the supernatant was collected and the protein amount was determined with the Micro BCA Protein Assay Kit (Thermofisher Scientific). For each condition, 2 mg of proteins were then pre-cleared with 15 μL of protein A/G sepharose beads (GE Healthcare) and placed in rotation at 4 °C at least 1 h. Following this pre-clearing stage, cells lysates were incubated with 4μg of SUZ12 antibody (ab12073, Abcam) or with 4 μg of Normal Rabbit IgG (#2729, Cell Signaling) on a rotator overnight at 4 °C. Thereafter, 20 μL of protein A/G sepharose beads were added and placed in rotation 1 h at 4 °C. After three washes with IPH buffer, bound proteins were eluted by boiling in 2X Laemmli buffer and resolved by SDS-PAGE.

4.13. Cut&Run Experiments

Protein-DNA interactions were analyzed using the CUT&RUN (Cleavage Under Targets & Release Using Nuclease) Assay Kit (#86652, Cell Signaling Technology) following the supplier's instructions. For each reaction, 250,000 cells were used. Positive (H3K4Me3) and negative (Rabbit mAb IgG) controls were performed using the antibodies provided in the kit. The EZH2 (#5246, Cell Signaling Technology) antibody was used at a dilution of 1/100. For DNA purification, the High Pure PCR Template Preparation Kit (#11796828001, Roche) was used according to the manufacturer's procedures except for the elution which was done in 50 μL buffer. Finally, binding of EZH2 and presence of the activating mark H3K4Me3 onto the *UNC5A* promoter was quantified by qPCR using the following primers: *UNC5A forward*: CCCTGACACCGTGTACATTCA, *UNC5A reverse*: TCACCACCTTCTGGTTTGGG.

5. Conclusions

In conclusion, we provided evidences that the nutritional sensor *O*-GlcNAcylation represents one of the molecular relays between nutritional disorders and the epigenetic downregulation of key tumor suppressor genes driving the progression of colorectal carcinogenesis including *UNC5A*. In the near future, it will therefore be interesting to investigate whether strategies targeting *O*-GlcNAcylation could represent a new therapeutic hope in the treatment of colorectal cancer.

Supplementary Materials: The following are available online at http://www.mdpi.com/2072-6694/12/11/3168/s1, Figure S1: Blood glucose levels; Figure S2: LS174T cells were maintained in EMEM (Lonza) supplemented with 10% fetal calf serum and 1% ZellShieldTM (Biovalley). Cells were cultured at 37 °C in water-saturated 5% CO_2 atmosphere. LS174T cells were transfected either with non-target siRNA control (siCTRL), OGT siRNA (siOGT), EZH2 siRNA (siEZH2) or with a combination of both for 72 h. Cells were harvested either for total proteins or RNA extraction; Table S1: RT-qPCR primers used in this study.

Author Contributions: Conceptualization: I.E.Y.-B. and V.D. Data curation: A.D., N.V., M.D., I.L., I.E.Y.-B. and V.D. Formal Analysis: A.D., N.V., M.D., I.L., J.T., E.L., O.C., D.L., I.E.Y.-B. and V.D. Funding acquisition: I.E.Y.-B. and V.D. Investigation: A.D., N.V., M.D., I.L. and J.T. Methodology: A.D., N.V., M.D., I.L., J.T., E.L., I.E.Y.-B. and V.D. Project Administration: I.E.Y.-B., V.D., O.C. and M.B.-M. Resources: I.E.Y.-B., V.D., E.L., O.C., M.B.-M. and D.L. Supervision: V.D. Validation: A.D., N.V., M.D., I.L., J.T., E.L., M.B.-M., O.C., D.L., I.E.Y.-B. and V.D. Visualization: A.D., I.L., M.D. and V.D. Writing—original draft: A.D., N.V., M.D., J.T., I.E.Y.-B. and V.D. All authors have read and agreed to the published version of the manuscript.

Funding: This work was supported by the "GEFLUC Flandres/Artois", the "Ligue Contre le Cancer/Comité du Nord", the "Cancéropôle Nord-Ouest" and the "Centre National de la Recherche Scientifique". A.D. is the recipient of a fellowship from the "Ministère de l'Enseignement Supérieur et de la Recherche".

Acknowledgments: We thank Andrea Paradisi and Patrick Mehlen (Cancer research center of Lyon, Lyon, France) for the gift of the *UNC5A* promoter plasmid. We thank Gerald Hart (Johns Hopkins University, Baltimore, Maryland, USA) and David Vocadlo (Simon Fraser University, Burnaby, Canada) for the gift of Ac5S-GlcNAc. This work has benefited from the facilities and expertise of the Platform for histology of Institute for translational research in inflammation (INFINITE U1286). This work was supported by the "GEFLUC Flandres/Artois", the "Ligue Contre le Cancer/Comité du Nord", the "Cancéropôle Nord-Ouest" and the "Centre National de la Recherche Scientifique". A.D. is the recipient of a fellowship from the "Ministère de l'Enseignement Supérieur et de la Recherche".

Conflicts of Interest: The authors declare no conflict of interest.

References

1. Bishop, K.S.; Ferguson, L.R. The interaction between epigenetics, nutrition and the development of cancer. *Nutrients* **2015**, *7*, 922–947. [CrossRef] [PubMed]
2. Sapienza, C.; Issa, J.-P. Diet, Nutrition, and Cancer Epigenetics. *Annu. Rev. Nutr.* **2016**, *36*, 665–681. [CrossRef] [PubMed]
3. Daniel, M.; Tollefsbol, T.O. Epigenetic linkage of aging, cancer and nutrition. *J. Exp. Biol.* **2015**, *218*, 59–70. [CrossRef]
4. Boyer, N.P.; Gupton, S.L. Revisiting Netrin-1: One Who Guides (Axons). *Front. Cell. Neurosci.* **2018**, *12*, 221. [CrossRef] [PubMed]
5. Mehlen, P.; Tauszig-Delamasure, S. Dependence receptors and colorectal cancer. *Gut* **2014**, *63*, 1821–1829. [CrossRef] [PubMed]
6. Shin, S.K.; Nagasaka, T.; Jung, B.H.; Matsubara, N.; Kim, W.H.; Carethers, J.M.; Boland, C.R.; Goel, A. Epigenetic and Genetic Alterations in Netrin-1 Receptors UNC5C and DCC in Human Colon Cancer. *Gastroenterology* **2007**, *133*, 1849–1857. [CrossRef] [PubMed]
7. Thiébault, K.; Mazelin, L.; Pays, L.; Llambi, F.; Joly, M.-O.; Scoazec, J.-Y.; Saurin, J.-C.; Romeo, G.; Mehlen, P. The netrin-1 receptors UNC5H are putative tumor suppressors controlling cell death commitment. *Proc. Natl. Acad. Sci. USA* **2003**, *100*, 4173–4178. [CrossRef]
8. Bernet, A.; Mazelin, L.; Coissieux, M.; Gadot, N.; Ackerman, S.L.; Scoazec, J.; Mehlen, P. Inactivation of the UNC5C Netrin-1 Receptor Is Associated With Tumor Progression in Colorectal Malignancies. *Gastroenterology* **2007**, *133*, 1840–1848. [CrossRef]

9. Hibi, K.; Mizukami, H.; Shirahata, A.; Goto, T.; Sakata, M.; Saito, M.; Ishibashi, K.; Kigawa, G.; Nemoto, H.; Sanada, Y. Aberrant methylation of the UNC5C gene is frequently detected in advanced colorectal cancer. *Anticancer. Res.* **2009**, *29*, 271–273.
10. Wu, J.; Wang, G.; He, B.; Chen, X.; An, Y. Methylation of the UNC5C gene and its protein expression in colorectal cancer. *Tumor Biol.* **2017**, *39*, 1010428317697564. [CrossRef]
11. Dehennaut, V.; Leprince, D.; Lefebvre, T. *O*-GlcNAcylation, an Epigenetic Mark. Focus on the Histone Code, TET Family Proteins, and Polycomb Group Proteins. *Front. Endocrinol.* **2014**, *5*, 155.
12. Decourcelle, A.; Leprince, D.; Edehennaut, V. Regulation of Polycomb Repression by O-GlcNAcylation: Linking Nutrition to Epigenetic Reprogramming in Embryonic Development and Cancer. *Front. Endocrinol.* **2019**, *10*, 117.
13. Leturcq, M.; Lefebvre, T.; Vercoutter-Edouart, A.-S. O-GlcNAcylation and chromatin remodeling in mammals: An up-to-date overview. *Biochem. Soc. Trans.* **2017**, *45*, 323–338. [PubMed]
14. Hardivillé, S.; Hart, G.W. Nutrient regulation of gene expression by *O*-GlcNAcylation of chromatin. *Curr. Opin. Chem. Biol.* **2016**, *33*, 88–94.
15. Zachara, N.; Akimoto, Y.; Hart, G.W. The O-GlcNAc Modification. In *Essentials of Glycobiology*; Varki, A., Ed.; Cold Spring Harbor Laboratory Press: New York, NY, USA, 2015.
16. Zheng, J.; Zhao, M.; Li, J.; Lou, G.; Yuan, Y.; Bu, S.; Xi, Y. Obesity-associated digestive cancers: A review of mechanisms and interventions. *Tumor Biol.* **2017**, *39*, 1010428317695020.
17. De Jong, R.G.; Peeters, P.J.; Burden, A.M.; De Bruin, M.L.; Haak, H.R.; Masclee, A.A.; De Vries, F.; Janssen-Heijnen, M.L. Gastrointestinal cancer incidence in type 2 diabetes mellitus; results from a large population-based cohort study in the UK. *Cancer Epidemiol.* **2018**, *54*, 104–111.
18. Pais, R.; Silaghi, H.; Silaghi, A.C.; Rusu, M.L.; Dumitrascu, D. Metabolic syndrome and risk of subsequent colorectal cancer. *World J. Gastroenterol.* **2009**, *15*, 5141–5148.
19. Olivier-Van Stichelen, S.; Dehennaut, V.; Buzy, A.; Zachayus, J.-L.; Guinez, C.; Mir, A.-M.; El Yazidi-Belkoura, I.; Copin, M.-C.; Boureme, D.; Loyaux, D.; et al. *O*-GlcNAcylation stabilizes β-catenin through direct competition with phosphorylation at threonine 41. *FASEB J.* **2014**, *28*, 3325–3338.
20. Mi, W.; Gu, Y.; Han, C.; Liu, H.; Fan, Q.; Zhang, X.; Cong, Q.; Yu, W. O-GlcNAcylation is a novel regulator of lung and colon cancer malignancy. *Biochim. Biophys. Acta (BBA)—Mol. Basis Dis.* **2011**, *1812*, 514–519.
21. Yang, Y.R.; Kim, D.H.; Seo, Y.-K.; Park, D.; Jang, H.-J.; Choi, S.Y.; Lee, Y.H.; Lee, G.H.; Nakajima, K.; Taniguchi, N.; et al. Elevated O-GlcNAcylation promotes colonic inflammation and tumorigenesis by modulating NF-κB signaling. *Oncotarget* **2015**, *6*, 12529–12542.
22. Vasconcelos-Dos-Santos, A.; Loponte, H.F.B.R.; Mantuano, N.R.; Oliveira, I.A.; De Paula, I.F.; Teixeira, L.K.; De-Freitas-Junior, J.C.M.; Gondim, K.C.; Heise, N.; Mohana-Borges, R.; et al. Hyperglycemia exacerbates colon cancer malignancy through hexosamine biosynthetic pathway. *Oncogenesis* **2017**, *6*, e306.
23. Xu, D.; Wang, W.; Bian, T.; Yang, W.; Shao, M.; Yang, H. Increased expression of O-GlcNAc transferase (OGT) is a biomarker for poor prognosis and allows tumorigenesis and invasion in colon cancer. *Int. J. Clin. Exp. Pathol.* **2019**, *12*, 1305–1314.
24. Steenackers, A.; Stichelen, S.O.-V.; Baldini, S.F.; Dehennaut, V.; Toillon, R.-A.; Le Bourhis, X.; El Yazidi-Belkoura, I.; Lefebvre, T. Silencing the Nucleocytoplasmic O-GlcNAc Transferase Reduces Proliferation, Adhesion, and Migration of Cancer and Fetal Human Colon Cell Lines. *Front. Endocrinol.* **2016**, *7*, 46.
25. Fardini, Y.; Dehennaut, V.; Lefebvre, T.; Issad, T. O-GlcNAcylation: A New Cancer Hallmark? *Front. Endocrinol.* **2013**, *4*, 99.
26. Fluge, Ø.; for the Norwegian Gastrointestinal Cancer Group; Gravdal, K.; Carlsen, E.; Vonen, B.; Kjellevold, K.; Refsum, S.; Lilleng, R.; Eide, T.; Halvorsen, T.; et al. Expression of EZH2 and Ki-67 in colorectal cancer and associations with treatment response and prognosis. *Br. J. Cancer* **2009**, *101*, 1282–1289. [PubMed]
27. Benard, A.; Goossens-Beumer, I.J.; Van Hoesel, A.Q.; Horati, H.; Putter, H.; Zeestraten, E.C.M.; Van De Velde, C.J.H.; Kuppen, P.J.K. Prognostic Value of Polycomb Proteins EZH2, BMI1 and SUZ12 and Histone Modification H3K27me3 in Colorectal Cancer. *PLoS ONE* **2014**, *9*, e108265.
28. Liu, Y.-L.; Gao, X.; Jiang, Y.; Zhang, G.; Sun, Z.-C.; Cui, B.-B.; Yang, Y.-M. Expression and clinicopathological significance of EED, SUZ12 and EZH2 mRNA in colorectal cancer. *J. Cancer Res. Clin. Oncol.* **2014**, *141*, 661–669.
29. Chu, C.-S.; Lo, P.-W.; Yeh, Y.-H.; Hsu, P.-H.; Peng, S.-H.; Teng, Y.-C.; Kang, M.-L.; Wong, C.-H.; Juan, L.-J. O-GlcNAcylation regulates EZH2 protein stability and function. *Proc. Natl. Acad. Sci. USA* **2014**, *111*, 1355–1360.

30. Inoue, D.; Fujino, T.; Sheridan, P.; Zhang, Y.-Z.; Nagase, R.; Horikawa, S.; Li, Z.; Matsui, H.; Kanai, A.; Saika, M.; et al. A novel ASXL1–OGT axis plays roles in H3K4 methylation and tumor suppression in myeloid malignancies. *Leukemia* **2018**, *32*, 1327–1337.
31. Jiang, M.; Xu, B.; Li, X.; Shang, Y.; Chu, Y.; Wang, W.; Chen, D.; Wu, N.; Hu, S.; Zhang, S.; et al. Correction: O-GlcNAcylation promotes colorectal cancer metastasis via the miR-101-O-GlcNAc/EZH2 regulatory feedback circuit. *Oncogene* **2019**, *38*, 5744–5745. [PubMed]
32. Forma, E.; Jóźwiak, P.; Ciesielski, P.; Zaczek, A.; Starska, K.; Bryś, M.; Krześlak, A. Impact of OGT deregulation on EZH2 target genes FOXA1 and FOXC1 expression in breast cancer cells. *PLoS ONE* **2018**, *13*, e0198351.
33. Shi, H.; Munk, A.; Nielsen, T.S.; Daughtry, M.R.; Larsson, L.; Li, S.; Høyer, K.F.; Geisler, H.W.; Sulek, K.; Kjøbsted, R.; et al. Skeletal muscle O-GlcNAc transferase is important for muscle energy homeostasis and whole-body insulin sensitivity. *Mol. Metab.* **2018**, *11*, 160–177. [PubMed]
34. Guo, H.; Zhang, B.; Nairn, A.V.; Nagy, T.; Moremen, K.W.; Buckhaults, P.; Pierce, M. O-LinkedN-Acetylglucosamine (O-GlcNAc) Expression Levels Epigenetically Regulate Colon Cancer Tumorigenesis by Affecting the Cancer Stem Cell Compartment via Modulating Expression of Transcriptional FactorMYBL1. *J. Biol. Chem.* **2017**, *292*, 4123–4137.
35. Neufert, C.; Becker, C.; Neurath, M.F. An inducible mouse model of colon carcinogenesis for the analysis of sporadic and inflammation-driven tumor progression. *Nat. Protoc.* **2007**, *2*, 1998–2004. [PubMed]
36. Plissonnier, M.-L.; Lahlali, T.; Raab, M.; Michelet, M.; Romero-López, C.; Rivoire, M.; Strebhardt, K.; Durantel, D.; Levrero, M.; Mehlen, P.; et al. Reciprocal antagonism between the netrin-1 receptor uncoordinated-phenotype-5A (UNC5A) and the hepatitis C virus. *Oncogene* **2017**, *36*, 6712–6724. [PubMed]
37. Barnault, R.; Lahlali, T.; Plissonnier, M.-L.; Romero-López, C.; Laverdure, N.; Ducarouge, B.; Rivoire, M.; Mehlen, P.; Zoulim, F.; Parent, R. Hepatocellular carcinoma-associated depletion of the netrin-1 receptor Uncoordinated Phenotype-5A (UNC5A) skews the hepatic unfolded protein response towards prosurvival outcomes. *Biochem. Biophys. Res. Commun.* **2018**, *495*, 2425–2431. [PubMed]
38. Lo, P.-W.; Shie, J.-J.; Chen, C.-H.; Wu, C.-Y.; Hsu, T.-L.; Wong, C.-H. O-GlcNAcylation regulates the stability and enzymatic activity of the histone methyltransferase EZH2. *Proc. Natl. Acad. Sci. USA* **2018**, *115*, 7302–7307.
39. Miyamoto, Y.; Futamura, M.; Kitamura, N.; Nakamura, Y.; Baba, H.; Arakawa, H. Identification of UNC5A as a novel transcriptional target of tumor suppressor p53 and a regulator of apoptosis. *Int. J. Oncol.* **2010**, *36*, 1253–1260.
40. Tanikawa, C.; Matsuda, K.; Fukuda, S.; Nakamura, Y.; Arakawa, H. p53RDL1 regulates p53-dependent apoptosis. *Nat. Cell Biol.* **2003**, *5*, 216–223.
41. Wang, H.; Ozaki, T.; Hossain, M.S.; Nakamura, Y.; Kamijo, T.; Xue, X.; Nakagawara, A. A newly identified dependence receptor UNC5H4 is induced during DNA damage-mediated apoptosis and transcriptional target of tumor suppressor p53. *Biochem. Biophys. Res. Commun.* **2008**, *370*, 594–598.
42. Zhu, Y.; Yu, M.; Chen, Y.; Wang, Y.; Wang, J.; Yang, C.; Bi, J. DNA damage-inducible gene, UNC5A, functions as a tumor-suppressor in bladder cancer. *Tumor Biol.* **2014**, *35*, 6887–6891.
43. Paradisi, A.; Creveaux, M.; Gibert, B.; Devailly, G.; Redoulez, E.; Neves, D.; Cleyssac, E.; Treilleux, I.; Klein, C.; Niederfellner, G.; et al. Combining chemotherapeutic agents and netrin-1 interference potentiates cancer cell death. *EMBO Mol. Med.* **2013**, *5*, 1821–1834.
44. Becker, C.; Fantini, M.C.; Wirtz, S.; Nikolaev, A.; Kiesslich, R.; A Lehr, H.; Galle, P.R.; Neurath, M.F. In vivo imaging of colitis and colon cancer development in mice using high resolution chromoendoscopy. *Gut* **2005**, *54*, 950–954. [PubMed]
45. Füzesi-Levi, M.G.; Ben-Nissan, G.; Bianchi, E.; Zhou, H.; Deery, M.J.; Lilley, K.S.; Levin, Y.; Sharon, M. Dynamic Regulation of the COP9 Signalosome in Response to DNA Damage. *Mol. Cell. Biol.* **2014**, *34*, 1066–1076.

Publisher's Note: MDPI stays neutral with regard to jurisdictional claims in published maps and institutional affiliations.

cancers

Article

Mitochondrial *O*-GlcNAc Transferase Interacts with and Modifies Many Proteins and Its Up-Regulation Affects Mitochondrial Function and Cellular Energy Homeostasis

Paweł Jóźwiak [1,*], Piotr Ciesielski [1], Piotr K. Zakrzewski [1], Karolina Kozal [1], Joanna Oracz [2], Grażyna Budryn [2], Dorota Żyżelewicz [2], Stéphanie Flament [3], Anne-Sophie Vercoutter-Edouart [4], Fabrice Bray [3], Tony Lefebvre [4] and Anna Krześlak [1]

[1] Department of Cytobiochemistry, Faculty of Biology and Environmental Protection, University of Lodz, 90-236 Lodz, Poland; piotr.ciesielski@biol.uni.lodz.pl (P.C.); piotr.zakrzewski@biol.uni.lodz.pl (P.K.Z.); karolina.kozal@edu.uni.lodz.pl (K.K.); anna.krzeslak@biol.uni.lodz.pl (A.K.)
[2] Institute of Food Technology and Analysis, Faculty of Biotechnology and Food Sciences, Lodz University of Technology, 90-924 Lodz, Poland; joanna.oracz@p.lodz.pl (J.O.); grazyna.budryn@p.lodz.pl (G.B.); dorota.zyzelewicz@p.lodz.pl (D.Ż.)
[3] Institut Eugène-Michel Chevreul, CNRS, MSAP USR 3290, FR 3688 FRABIO, FR 2638, Université de Lille, 59000 Lille, France; stephanie.flament@univ-lille.fr (S.F.); fabrice.bray@univ-lille.fr (F.B.)
[4] Unité de Glycobiologie Structurale et Fonctionnelle, CNRS, UMR 8576, UGSF, Université de Lille, 59000 Lille, France; anne-sophie.vercoutter@univ-lille.fr (A.-S.V.-E.); tony.lefebvre@univ-lille.fr (T.L.)
* Correspondence: pawel.jozwiak@biol.uni.lodz.pl; Tel.: +48-42-6354403

Citation: Jóźwiak, P.; Ciesielski, P.; Zakrzewski, P.K.; Kozal, K.; Oracz, J.; Budryn, G.; Żyżelewicz, D.; Flament, S.; Vercoutter-Edouart, A.-S.; Bray, F.; et al. Mitochondrial *O*-GlcNAc Transferase Interacts with and Modifies Many Proteins and Its Up-Regulation Affects Mitochondrial Function and Cellular Energy Homeostasis. *Cancers* **2021**, *13*, 2956. https://doi.org/10.3390/cancers13122956

Academic Editor: Marcelo G. Kazanietz

Received: 17 May 2021
Accepted: 8 June 2021
Published: 12 June 2021

Simple Summary: *O*-GlcNAcylation is a dynamic post-translational modification of proteins involved in the control of intracellular signaling pathways in response to changes in nutrient availability, especially glucose concentration. To date, most research has focused on *O*-GlcNAcylation of proteins by the nuclear-cytoplasmic isoform of *O*-GlcNAc transferase (ncOGT), while the role of mitochondrial OGT (mOGT) and its effect on *O*-GlcNAcylation of mitochondrial proteins are poorly understood. The aim of our study was to investigate the effect of mOGT on *O*-GlcNAcylation of mitochondrial proteins, mitochondrial function, and energy metabolism of breast cancer cells. Herein, we used two independent proteomics-based approaches to identify mOGT-interacting partners and proteins modified by mOGT. Based on our findings, we propose that *O*-GlcNAcylation of proteins by mOGT is a part of the mechanism by which glucose affects mitochondrial function and cellular bioenergetics.

Abstract: *O*-GlcNAcylation is a cell glucose sensor. The addition of *O*-GlcNAc moieties to target protein is catalyzed by the *O*-Linked *N*-acetylglucosamine transferase (OGT). OGT is encoded by a single gene that yields differentially spliced OGT isoforms. One of them is targeted to mitochondria (mOGT). Although the impact of *O*-GlcNAcylation on cancer cells biology is well documented, mOGT's role remains poorly investigated. We performed studies using breast cancer cells with up-regulated mOGT or its catalytic inactive mutant to identify proteins specifically modified by mOGT. Proteomic approaches included isolation of mOGT protein partners and *O*-GlcNAcylated proteins from mitochondria-enriched fraction followed by their analysis by mass spectrometry. Moreover, we analyzed the impact of mOGT dysregulation on mitochondrial activity and cellular metabolism using a variety of biochemical assays. We found that mitochondrial OGT expression is glucose-dependent. Elevated mOGT expression affected the mitochondrial transmembrane potential and increased intramitochondrial ROS generation. Moreover, mOGT up-regulation caused a decrease in cellular ATP level. We identified many mitochondrial proteins as mOGT substrates. Most of these proteins are localized in the mitochondrial matrix and the inner mitochondrial membrane and participate in mitochondrial respiration, fatty acid metabolism, transport, translation, apoptosis, and mtDNA processes. Our findings suggest that mOGT interacts with and modifies many mitochondrial proteins, and its dysregulation affects cellular bioenergetics and mitochondria function.

Keywords: mOGT; *O*-GlcNAc; mitochondria; glucose; energy metabolism; breast cancer

1. Introduction

Cancer cells undergo significant metabolic changes that allow rapid growth and proliferation. The classic bioenergetic phenotype of cancer cells usually displays an enhanced rate of glycolysis along with inhibition of oxidative phosphorylation even in the presence of oxygen, a phenomenon known as "Warburg effect". It is well-known that these metabolic alterations are strongly dependent on glucose consumption and utilization [1,2].

Within the cell, glucose is mainly metabolized into pyruvate through glycolysis, but a fraction of approximately 2–4% enters the hexosamine biosynthetic pathway (HBP). The end product of the HBP is the nucleotide sugar uridine diphosphate N-acetylglucosamine (UDP-GlcNAc), used for many reactions as a sugar donor, including β-*O*-linked N-acetylglucosaminylation (*O*-GlcNAcylation) [3]. All *O*-GlcNAc-modified proteins are phosphoproteins and sometimes *O*-GlcNAc and *O*-phosphate moieties compete for the same or neighboring site [4]. The interplay between the two post-translational modifications can determine target protein function such as subcellular transport and localization, complex formation, or enzyme activity; therefore, the reciprocal relationship between these modifications plays a significant role in the regulation of multiple biological processes [5]. Dynamic *O*-GlcNAc cycling is regulated by two key enzymes: the *O*-linked β-N-acetyl-D-glucosaminyltransferase (OGT) is responsible for the addition of a single GlcNAc residue to proteins, whereas *O*-linked β-N-acetyl-D-glucosaminidase (OGA) catalyzes the removal of the GlcNAc moiety [6,7]. In humans, there is only a single *OGT* gene located on chromosome Xq13.1, a region associated with Parkinson's disease [8]. To date, three functional isoforms of the OGT enzyme, arising possibly by alternative splicing or alternative promoters, have been identified. The longest (~116 kDa, ncOGT) and the shortest (~78 kDa, sOGT) isoforms are located in both the nucleus and the cytoplasm (Figure 1A). The third isoform (~103 kDa) contains a mitochondrial targeting sequence (MTS) at its N-terminus and has been referred to as mitochondrial OGT (mOGT) [9].

Over the last two decades, there has been solid evidence supporting a pivotal role of *O*-GlcNAcylation in cancer cell biology. The results of many studies suggest that increased expression of OGT and hyper-*O*-GlcNAcylation are the universal features of cancers, including breast cancer [10,11]. Accumulating evidence suggests that OGT may act as a nutrient sensor that links HBP to oncogenic signaling and regulation of factors involved in glucose metabolism and cancer progression. *O*-GlcNAc signaling has been proven essential for the onset, progression, and metastasis of breast cancer [12,13]. It has been shown that decreased *O*-GlcNAcylation as a result of reduced expression of OGT by siRNA causes profound changes in the proteome of MCF-7 breast cancer cells. The gel-free quantitative proteomics coupled with LC–MS/MS analysis allows identifying proteins affected by *O*-GlcNAc inhibition, which may have important roles in cancer metastasis. Among proteins whose expression was affected by *O*-GlcNAc changes, proteins involved in cellular metabolism, cellular localization, stress responses, and gene expression were identified [14]. In spite of *O*-GlcNAc significance in breast cancer cells, the impact of *O*-GlcNAc cycling enzymes on mitochondria function was not yet investigated. However, there are several examples of roles of *O*-GlcNAc in other cancerous and non-cancerous cell types. Elevated *O*-GlcNAcylation of respiratory chain complex I and III proteins is associated with impaired mitochondrial function in cardiomyocytes under hyperglycemic conditions [15]. Furthermore, *O*-GlcNAc cycling modulates activity and translocation of the dynamin-related protein 1 (DRP1), whose activation is associated with the loss of membrane potential and mitochondria fragmentation [16]. A recent study performed on rat purified cardiac mitochondria revealed that among the 88 identified *O*-GlcNAcylated mitochondrial candidate proteins, nearly half of them are components of the oxidative phosphorylation system [17]. In addition, the identification of the pyrimidine nucleotide carrier 1 (pnc1) as a UDP-GlcNAc transporter reinforces the view that mitochondria possess the machinery for switching *O*-GlcNAc cycling on and off [18]. Until now, the studies concerning the links between *O*-GlcNAcylation and cancer biology have been concentrated on the ncOGT activity, whereas the reports related to the role of mOGT are very limited.

Trapannone et al. [19] suggested that ncOGT was sufficient for *O*-GlcNAc moieties attachment to mitochondrial proteins. However, the identification of some *O*-GlcNAcylated proteins encoded by mitochondrial DNA (cytochrome oxidase 1; MTCO1; cytochrome oxidase 2; COX2 and NADH:ubiquinone oxidoreductase core subunit 4; MT-ND4) raises the possibility that the mOGT isoform is required for their modification. Depletion of mOGT by siRNA allowed the identification of four putative mOGT's protein substrates [15–17]. The vast majority of mitochondrial proteins are encoded by nuclear genes and transported into mitochondria after synthesis in the cytosol. Therefore, part of proteins residing in mitochondria could be *O*-GlcNAcylated by the ncOGT prior to translocation to mitochondria. However, early reports clearly showed that some proteins, such as Nup62 and casein kinase II, were glycosylated by both ncOGT and mOGT, while other ones, such as yes kinase, were specifically modified by mOGT [20]. The substrate selectivity most likely occurs due to the different number of tetratricopeptide repeats (TPR) in the N-terminal region of OGT isoforms [21]. Therefore, mOGT may have different effects on mitochondria function than ncOGT; as an example, mOGT, overexpressed in cancer cell lines, could induce apoptosis probably as a pro-apoptotic partner. However, the impact of mOGT on mitochondrial homeostasis and cellular physiology still remains substantially unknown.

Thus, in this study, in order to bring closer the significance of mOGT in cancer cells, we identified mOGT protein substrates and investigated the effect of mOGT dysregulation on mitochondrial activity of breast cancer cells.

2. Materials and Methods

2.1. Reagents and Antibodies

Chemicals were obtained from Sigma-Aldrich (St. Louis, MO, USA) except when specified. Cell culture reagents and materials were purchased from Invitrogen (Carlsbad, CA, USA), Cytogen (Sinn, Germany), and Corning Inc. (Corning, NY, USA). The used antibodies—mouse monoclonal anti-cytochrome C, mouse monoclonal anti-β-actin, and mouse monoclonal anti-lamin A/C—were from Santa Cruz Biotechnology, Inc. (Santa Cruz, CA, USA). The monoclonal mouse anti-*O*-GlcNAc (RL2) (ab2739), monoclonal mouse anti-BrdU (IIB5) (ab8152), and polyclonal goat anti-mouse IgG H&L FITC (ab6785) antibodies were from Abcam (Cambridge, UK). Rabbit polyclonal anti-*O*-GlcNAcase (OGA) (SAB4200311) and rabbit polyclonal anti-*O*-GlcNAc transferase (OGT) (Ti-14) antibodies were from Sigma-Aldrich (St. Louis, MO, USA). Rabbit monoclonal anti-OGT (G921A), secondary mouse anti-rabbit (7074), and goat anti-mouse (7076) IgG-HRP antibodies were purchased from Cell Signaling Technology, Inc. (Beverly, MA, USA). Mouse monoclonal anti-HaloTag antibodies were from Promega® (Madison, WI, USA).

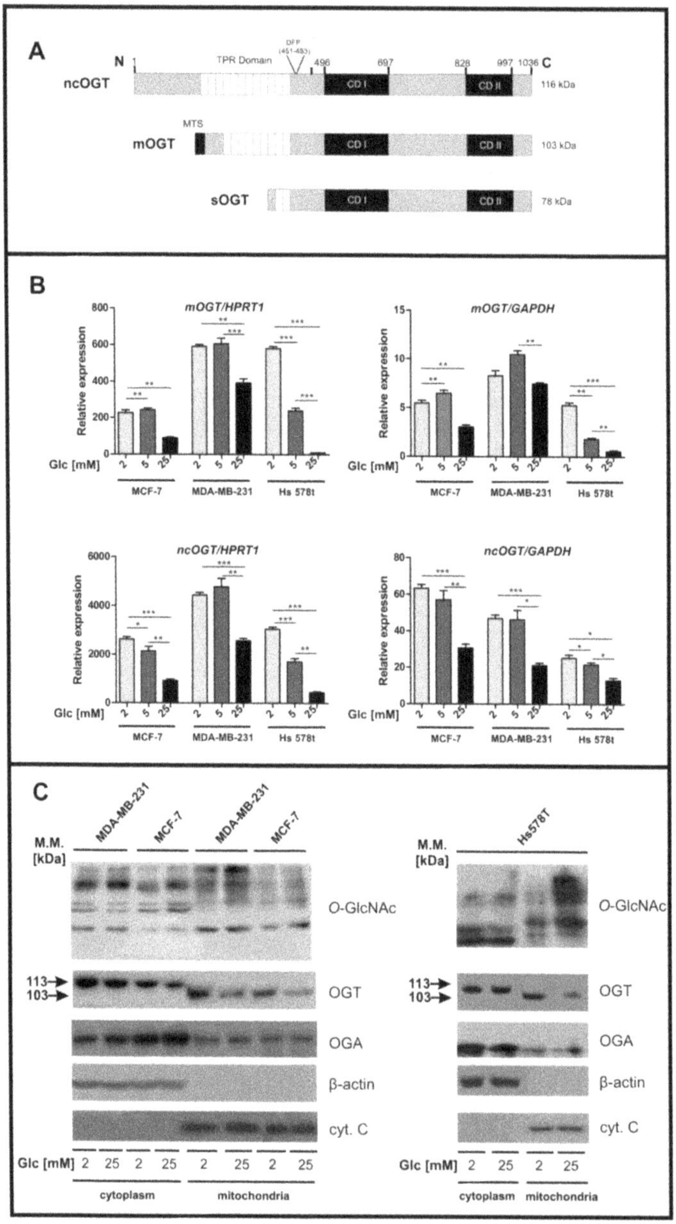

Figure 1. Glucose concentration alters the expression of OGT isoforms and *O*-GlcNAc levels in mitochondria. (**A**) Schematic structure of OGT isoforms. (**B**) *mOGT* and *ncOGT* transcript levels were determined using RT-qPCR, and the results were referenced to *HPRT1* and *GAPDH* housekeeping genes mRNA. The bars represent mean +/− SD from three independent experiments in duplicate. (**C**) *O*-GlcNAc, OGT, and OGA levels in cytoplasmic (30 μg of proteins per line) and mitochondrial (60 μg of proteins per line) fractions of breast cancer cells. For cytoplasmic extract, β-actin was used as a loading control, whereas cytochrome C was used as a marker of mitochondria-enriched fraction. * indicates significance; $p < 0.05$; ** $p < 0.01$; *** $p < 0.001$. Glc, glucose. (The original western blot images are shown in the File S1).

2.2. DNA Constructs

The full-length human mOGT gene reported in GenBank™ (accession number U77413) was synthesized using GeneArt service (Invitrogen™, Carlsbad, CA, USA). The ΔCD-mOGT catalytically inactive mOGT mutant was generated using the sequence encoding mOGT but without the last 93 amino acid region (Figure 2A).

The codon sequences were normalized by GeneOptimizer™ software (Thermo Fisher Scientific, Waltham, MA, USA) in order to obtain a high yield of mRNAs and proteins from synthetic genes. Each synthesized construct was flanked by SgfI and PmeI sites in pMA vector and subcloned into the pFC27K HaloTag® CMV-neo Flexi® Vector (#G8431, Promega™) following the manufacturer's instructions and transformed into competent JM109 cells (Promega™, Madison, WI, USA). These CMV-driven constructs produce fusion proteins tagged by the HaloTag protein, which consists of a 33 kDa monomeric protein not expressed in mammalian, plant, or *E. coli* cells. The empty vector that only encodes HaloTag protein was generated by blunt ends ligation of purified product generated after PCR amplification using designed primers and the Physion™ (Thermo Scientific™, Waltham, MA, USA) High-Fidelity DNA Polymerase (Thermo Scientific™). Plasmids were isolated using Extract Me Plasmid Maxi Endotoxin-Free Kit (Blirt®, Gdańsk, Poland) and validated by direct DNA sequencing.

2.3. Cell Culture and Treatment

MCF-7, MDA-MB-231, and Hs578t breast cancer cell lines were obtained from the American Type Culture Collection (Manassas, VA, USA). Cells were grown in Dulbecco's modified Eagle's medium (DMEM) supplemented with 10% (v/v) fetal bovine serum (FBS). All cell lines were grown in a humidified atmosphere containing 5% (v/v) CO_2 at 37 °C. To assess the impact of glucose on levels of *O*-GlcNAcylation as well as expression of proteins, cells were grown for 72 h in medium containing 2, 5, or 25 mM glucose concentrations, which in blood correspond to hypo-, normo-, and hyperglycemia conditions, respectively. Transfections with vectors encoding mOGT, ΔCD-mOGT, or HaloTag protein alone were performed using Lipofectamine™ 2000 (Invitrogen™, ThermoFisher Scientific, Grand Island, NY, USA) transfection reagent. Cells were seeded on plates at 90% confluence and then transfected for 24 or 48 h with 0.1 μg of plasmid and 0.2 μL Lipofectamine per well (96-well plate) or 2 μg of pDNA and 4 μL of Lipofectamine per well (6-well plate) according to the manufacturer's instructions. For experiments dedicated to the identification of mOGT's protein substrates and binding partners by mass spectrometry, cells were co-transfected with plasmids in combination with Silencer™ Pre-designed siRNA targeted ncOGT or with Silencer™ Negative Control No.1 siRNAs (Cat. Nr. AM16704 and AM4611, respectively; Ambion®, (Milwaukee, WI, USA). Initially, cells plated on 10 cm dishes at 70% confluence were transfected for 72 h with 100 pmol siRNA in the presence of 10 μL Lipofectamine RNAiMAX (Invitrogen™, ThermoFisher Scientific, Grand Island, NY, USA) following manufacturer's specifications. After 24 h, cells were co-transfected with 10 μg of plasmid using Lipofectamine 2000 transfection reagent. Cells were then incubated for 48 h. For all experiments, cells were plated in triplicates.

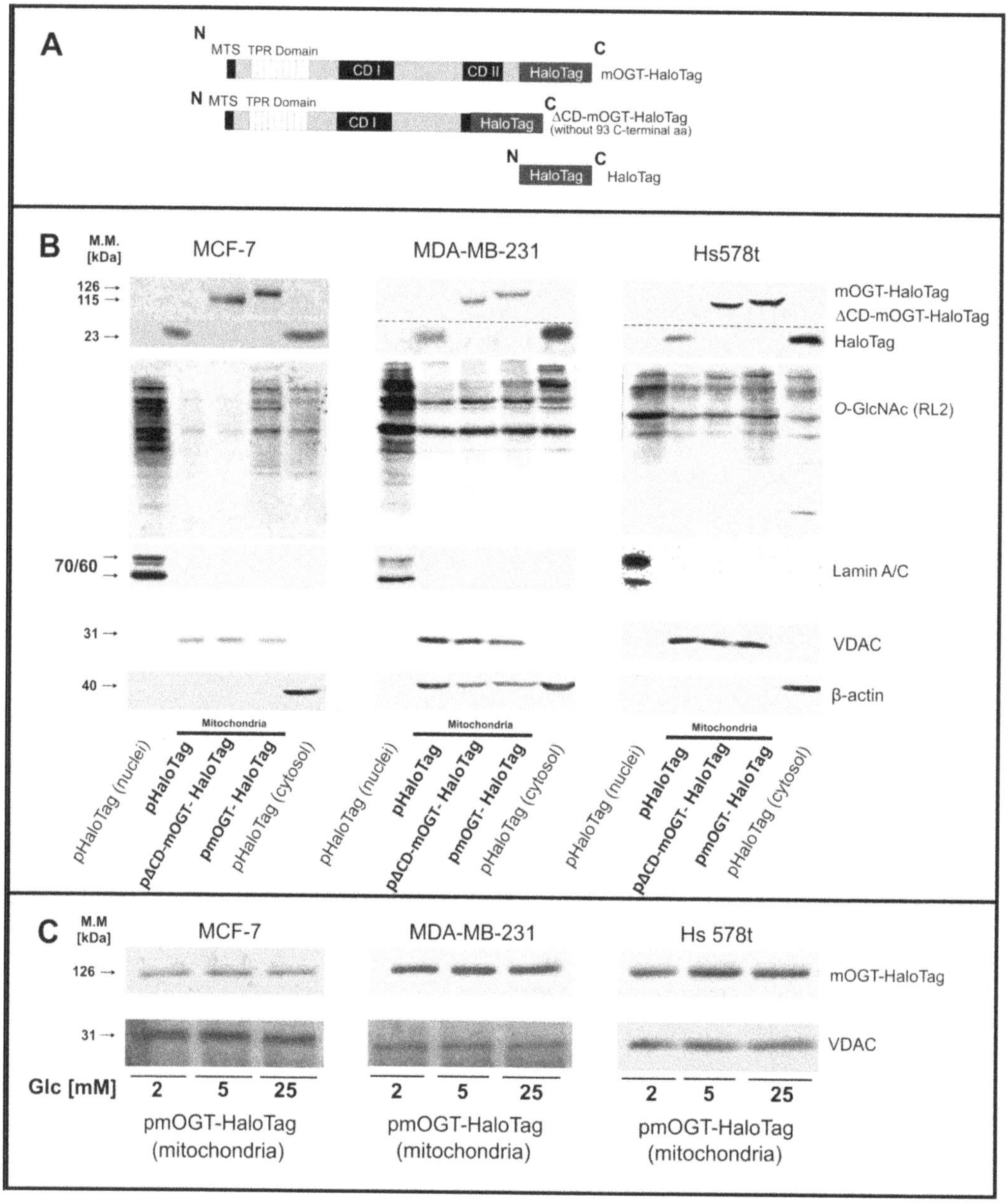

Figure 2. Up-regulated mOGT targets the mitochondria and modifies mitochondrial proteins in breast cancer cells. (**A**) The scheme of mOGT-HaloTag fused protein, its catalytic inactive mutant with truncated CDII domain, and HaloTag used as a control. (**B**) Expression of Halo-tagged proteins and *O*-GlcNAc level in mitochondria-enriched fractions 48 h after plasmid transfection of cells growing in hypoglycemia. Nuclei-enriched fraction and cytoplasm extract were loaded to assess the purity of mitochondria. VDAC, Lamin A/C, and β-actin were used as purity markers for mitochondria, nucleus, and cytoplasm, respectively. (**C**) Expression of mOGT-HaloTag in mitochondria-enriched fraction derived from cells cultured in hypo-, normo-, or hyperglycemia conditions. MTS, mitochondria-targeted sequence; TPR, tetratricopeptide repeats, CD, catalytic domain; Glc, glucose. (The original western blot images are shown in the File S1).

2.4. RT-PCR

RNA was isolated from the breast cancer cells using the total RNA isolation kit (A&A Biotechnology, Gdynia, Poland) according to the manufacturer's instructions. First-strand cDNAs were obtained by the reverse transcription of 1 μg of total RNA using a High-Capacity cDNA Reverse Transcription kit (Applied Biosystems, Foster City, CA, USA); then, qPCR was performed using HOT FIREPol®EvaGreen® qPCR Mix Plus (Solis Bio-Dyne, Tartu, Estonia) according to the manufacturer's instructions. The following sets of primers were employed to measure the expression levels of mOGT and ncOGT and the internal controls HPRT1 or GAPDH by qPCR: mOGT, 5′-tggctggtcagagaaggaataa-3′, and 3′-gacgttggatcggttacgag-5′; ncOGT, 5′-gcaacctagccaatgctctc-3′, and 3′-cagaagggtctcaaacgacg-5; HPRT1 5′-ccctggcgtcgtgattagtg-3′ and 3′-cctgacttgcagaacgagct-5′; GAPDH 5′-cctgcaccaccaactgctta-3′ and 3′-aggtcttgtagtagggacgg-5′. Detailed information concerning the primers is presented in File S2 Supplementary Methods. The reactions were performed in triplicate using Eppendorf RealPlex thermal cycler (Enfield, CT, USA). PCR products were confirmed by melting curves analysis and were tested by electrophoresis with ethydium bromide staining. The abundance of studied genes mRNA in samples was quantified by the ΔC method. Ct (Ct_{gene}–Ct_{GAPDH}); values were recalculated into relative copy number values (number of copies of gene of interest mRNA per 1000 copies of reference gene mRNA).

2.5. Isolation of Mitochondrial, Nucleus, and Cytosol Fractions

Mitochondria were purified from cells using Mitochondria Isolation Kit for Cultured Cells (Thermo Scientific) according to manufacturer's instructions, with minor modifications. Unless specified otherwise, all steps were carried out at 4 °C. A total of 2×10^7 cells were briefly suspended in Mitochondria Isolation Reagent A and incubated for 2 min on ice. Cells were transferred into a Glass-Teflon homogenizer and disrupted by ~25–30 strokes. The lysis efficiency was determined by visual estimation using a microscope. Subsequently, Mitochondria Isolation Reagent C was added, and then lysed cells were centrifuged at 700× *g* for 10 min to remove cell debris and remaining nuclei. For SDS-PAGE, pellets containing nuclei were purified three times by suspension in Reagent C followed by centrifugation at 700× *g* for 10 min. The lacking nuclei supernatants were also spun again at 700× *g* for 10 min to remove the remaining nuclei. Then, supernatants were transferred into new vials and spun at 3000× *g* for 15 min. The pellets containing mitochondria were resuspended in Mitochondria Isolation Reagent C and centrifuged at 12,000× *g* for 10 min. Mitochondria-enriched pellets were collected and used for subsequent experiments. Cytosol fractions were purified by twice centrifugation at 12000× *g* for 10 min of supernatants collected following mitochondria sedimentation.

2.6. Western Blotting

Proteins were resolved by SDS-PAGE and electroblotted onto Immobilon-P transfer membranes. The blots were incubated with primary antibodies for 2 h at room temperature. After washing three times with Tris-buffered saline (TBS), the blots were incubated for 1 h with horseradish peroxidase-coupled secondary antibodies. Proteins were visualized on X-ray films or by using a CCD camera by an enhanced chemiluminescence method using SuperSignal™ West Pico PLUS Chemiluminescent Substrate (Thermo Scientific™). For loading control, the blots were re-probed with anti-β-actin, anti-cytochrome C, anti-VDAC1, anti-VDAC3, or anti-lamin A/C antibodies following a stripping protocol.

2.7. Determination of Mitochondrial ROS Production

Mitochondrial ROS production was evaluated using MitoSOX assay. Cells were seeded onto 12-well plates and incubated overnight. Cells were treated for 24 or 48 h with plasmid DNA followed by staining with MitoSOX Red mitochondrial superoxide indicator (Molecular Probes, Eugene, OR, USA). After washing of the cells, the specific localization of the dye was confirmed using a fluorescence microscope (Olympus IX70,

Melville, NY, USA)) and then the fluorescence intensity of the cells was assessed by flow cytometry (excitation wavelength: 510 nm; emission wavelength: 580 nm; cytometer Becton DickinsonLSR®, (Bergen, NJ, USA).

2.8. ATP Level Measurement

ATP level was determined by using a Luminescent ATP Detection Assay Kit (ID: ab113849; Abcam®, Cambridge, UK) per the manufacturer's instruction. Briefly, cells were seeded in a 96-well plate (μClear®, white, flat bottom; Greiner Bio-One GmbH, Frickenhausen, Germany) one day prior to the treatment. Cells growing in media with different glucose concentrations (low, medium, and high) were transfected with plasmid DNA for 24 or 48 h, and then ATP level was measured. Cell lysates were treated with ATP detection reagent for 10 min at room temperature. The luminescent signal was quantified by a plate reader (SpectraMax® PARADIGM® Multimode Microplate Plate Reader; Molecular Devices, CA, USA). Cellular ATP levels were calculated from a standard curve (μM).

2.9. Mitochondrial Membrane Potential Assay

Mitochondrial membrane potential was assessed in cells cultured in 96-well plate (μClear®, black, flat bottom; Greiner Bio-One GmbH, Frickenhausen, Germany). Cells were stained with cell-permeable potentiometric fluorescent dye JC-1 (5,5′,6,6′-tetrachloro-1,1′,3,3′-tetraethylbenzimidazolycarbocyanine iodide) as described previously [22]. Cells were labeled with JC-1 at a concentration of 2 μM and incubated at 37 °C for 30 min before being washed once with pre-warmed (37 °C) PBS. For the positive control, cells were incubated with 50 μM final concentration of CCCP before staining. The fluorescence was measured on a Bio-Tek Synergy HT Microplate Reader (Bio-Tek Instruments, Winooski, VT, USA), with filter pairs of 485 nm/538 nm and 530 nm/590 nm. Results were shown as a ratio of fluorescence, measured at 530 nm/590 nm to that measured at 485 nm/538 nm (aggregates to monomer fluorescence).

2.10. Cell Imaging

For experiments with RNA interference, cells were plated at a density of 1.5×10^4 cells/well for 24 h and then treated with siRNA duplexes for 72 h, as described in the cell culture and treatment section. Following transfection, cells were fixed in 4% (m/v) paraformaldehyde, washed in PBS, quenched in 100 mM glycine, washed in PBS, and permeabilized in PBS containing 0.5% (v/v) Triton X-100 for 20 min. Then, cells were blocked in blocking buffer (2% (v/v) Fetal Calf Serum (FCS)-2% (m/v) Bovine Serum Albumin (BSA)-0.2% (m/v) gelatin) for 1 h at room temperature. Cells were immunostained for OGT (Ti-14, Sigma-Aldrich) at a 1:100 dilution in blocking solution overnight at 4 °C. After washing in PBS, cells were incubated with the Alexa Fluor 647-conjugated goat anti-rabbit IgG antibodies (1:600; Invitrogen) for 1 h at room temperature. Cells were then washed with PBS and counterstained with DAPI (50 μg/mL, Molecular Probes) to visualize nuclei. Immunolabeled cells were imaged at 25 °C at a magnification of 63× by using a Leica TCS SP8 system.

2.11. Enrichment of Mitochondrial O-GlcNAcylated Proteins and Isolation of mOGT-Interacting Protein Partners by Immobilization of HaloTag® Fusion Proteins

Immunoprecipitation was performed using RL2 antibody on mitochondria-enriched fractions from cells co-transfected with plasmids (empty backbone; ΔCD-mOGT-HaloTag; mOGT-HaloTag) and ncOGT siRNA. Immunoprecipitates were carried out using Pierce® Crosslink Immunoprecipitation Kit (Thermo Scientific, Rockford, IL, USA) according to the manufacturer's instruction. Mitochondria-enriched samples were incubated for 30 min in a lysis buffer (50 mM Tris-HCl, 150 mM NaCl, 1% (v/v) NP-40, 0.5% (m/v) sodium deoxycholate, 0.1% (m/v) SDS, pH 7.4, supplemented with Halt™ protease Inhibitor Cocktail (Thermo Scientific, Rockford, IL, USA) on ice. The lysates were mixed at a volume ratio 1:9 with Pierce® IP Lysis buffer (25 mM Tris-HCl, 150 mM NaCl, 1 mM EDTA, 1%

(*v*/*v*) NP-40, 5% (*v*/*v*) glycerol, pH 7.4, supplemented with Halt™ Protease Inhibitor Cocktail (Thermo Scientific, Rockford, IL, USA) and then centrifuged at 13,000× *g* for 10 min. Supernatants (0.5–0.7 mg of protein) were pre-cleared using control agarose resin. Immunoprecipitation experiments were carried out by incubating pre-cleared lysates with antibody-crosslinked resin at 4 °C with gentle rolling overnight. After four washes, the beads were incubated with Laemmli sample buffer containing 5% (*v*/*v*) 2-mercaptoethanol and heated at 95 °C for 5 min.

Isolation of protein partners interacting with mOGT-HaloTag, ΔCD-mOGT-HaloTag fusion proteins, or HaloTag protein only were done by pull-down assay using HaloLink™ Resin (Promega™, Madison, WI, USA) according to manufacturer's instructions. For this method, HaloTag binding ligand was covalently attached to the surface of Sepharose beads via a carbamide linkage. Covalent binding of HaloTag® fusion proteins to the HaloLink™ Resin allows extensive washing of the bait-prey complexes without the risk of dissociation from the resin. About 500 µg of proteins derived from mitochondria-enriched lysates of cells transfected with plasmid DNA was added to 150 µL equilibrated HaloLink™Resin resuspended in binding buffer (100 mM Tris-HCl, 150 mM NaCl, 0.05 % (*v*/*v*) IGEPAL® CA-630, pH 7.6) and then incubated by mixing overnight on a tube rotator at 4°C. The precipitates were washed four times with 1 mL of buffer containing 100 mM Tris-HCl, −150 mM NaCl, 1 mg/mL BSA 0.5%, IGEPAL® CA-630 pH 7.6) followed by centrifugation at 800× *g* for 2 min. Finally, proteins were eluted by heating at 95 °C with a Laemmli sample loading buffer (2×).

2.12. *Identification of mOGT's Protein Partners and Protein Substrates by Mass Spectrometry*

2.12.1. Chemicals

Pierce® LTQ ESI positive ion calibration solution and Pierce™C18 tips were purchased from Thermo Fisher Scientific (Rockford, IL, USA). LC–MS-grade acetonitrile, HPLC-grade formic acid and trifluoroacetic acid, MS Qual/Quant QC Mix, were obtained from Sigma-Aldrich (St. Louis, MO, USA). Water was purified by a Milli-Q water purification system (Millipore Corp., Bedford, MA, USA).

2.12.2. Sample Preparation Prior to LC–MS/MS

The eluted proteins were resolved by SDS-PAGE, and gels were stained using brilliant blue staining. Each lane was cut into small slices and placed into 0.65 mL siliconized tubes. The slices were washed thrice in ~100 µL of 25 mM NH_4HCO_3/50% (*v*/*v*) ACN by vortexing for 10 min followed by complete drying of the gel pieces using a Speed Vac centrifuge (Eppendorf). The dried gels were incubated with 10 mM DTT in 25 mM NH_4HCO_3 at 56 °C for 1 h. Supernatants were removed, and the gel pieces were covered by 55 mM iodoacetamide in 25 mM NH_4HCO_3 and were placed in the dark for 45 min at room temperature. After this step, gels were washed by vortexing for 10 min with ~100 µL NH_4HCO_3, then shaken twice with ~100 µL of 25 mM NH_4NCO_3 in 50% (*v*/*v*) ACN for 5 min. Dehydrated gels were centrifuged to complete dryness using a Speed Vac centrifuge (Eppendorf). Next, the gel pieces were rehydrated at 4 °C with 25 µL of Trypsin Gold (Promega, Madison, WI, USA) solution (12.5 ng/µL trypsin in 25 mM NH_4CO_3, freshly diluted). As needed, a solution of 25 mM NH_4HCO_3 was added to cover the slices. Digestion was performed overnight at 37 °C. The digest solution was transferred into a clean 0.5 mL siliconized tube. The gel pieces were mixed twice with 50 µL of 50% (*v*/*v*) ACN/5% (*v*/*v*) formic acid. The digests were dried using a Speed Vac centrifuge, then re-suspended in 0.1% (*v*/*v*) trifluoroacetic acid (TFA) in water and purified by the Pierce™ C18 tips according to the manual instructions. Briefly, C18 tips were activated with 100 µL acetonitrile (e.g., 8–10 up–down cycles when attached to a 10 µL pipette) and stabilized by 50% (*v*/*v*) acetonitrile and 0.1% (*v*/*v*) TFA in water. Next, 10 µL of sample was loaded into the Pierce™ C18 tip (with 10 up–down cycles of the pipette). Then, the trapped peptide sample was washed with 100 µL 0.1% (*v*/*v*) TFA. Finally, the sample was eluted from the

Pierce™ C18 tip using 20 μL of 0.1% (v/v) formic acid in a 95% (v/v) acetonitrile into a vial for LC–MS/MS analysis.

2.12.3. LC–MS/MS Analysis

LC–MS/MS analysis was performed using a Transcend™ TLX-2 multiplexed LC system equipped with Q-Exactive Orbitrap mass spectrometer (Thermo Scientific, Hudson, New Hampshire, USA) using a heated electrospray ionization (ESI) interface (HESI–II) according to the method described by Kockmann et al. [23], Geiger et al. [24], and Velloso et al. [25], with some modifications. Samples were separated using an aC18 Acclaim PepMap™ 100 column (1.0 × 150 mm, 3 μm particle size, nanoViper, Thermo Fisher Scientific, PA, USA) thermostated at 25 °C. The mobile phases were eluent A, FA/water (0.1/99.9, v/v) and eluent B, FA/acetonitrile (0.1/99.9, v/v). The flow rate was 75 μL/min, and the gradient was as follows: 0–0.25 min, 2% B; 0.25–54 min, 2–35% B; 54–75 min, 35–2% B; 2% B was maintained for another 15 min for column re-equilibration. The sample injection volume was 10 μL. The Q-Exactive Orbitrap mass spectrometer was operated in a positive ionization mode with full MS and a subsequent all-ion fragmentation (AIF) mode. The accuracy and mass calibration were done according to the manufacturer's recommendations using a mixture of standards in the mass range of m/z 138.06619–1779.96528. The capillary temperature was adjusted to 250 °C and aux gas heater temperature to 100 °C. The electrospray capillary voltage and S-lens radio frequency (RF) level were set at 2.5 kV and 50 V, respectively. Nitrogen was used as both sheath gas and auxiliary gas at a flow of 5 and 1 (arbitrary units), respectively. The acquisition method consisted of two scan events, full MS-SIM and AIF. The full MS-SIM scan spectra were acquired over an m/z range of 350–2000, with the mass resolution of 17,000 full-width at half maximum (FWHM) at m/z = 200. Automatic gain control (AGC) target (the number of ions to fill C-Trap) was set at 3.0e^6 with a maximum injection time (IT) of 50 ms. The second scan event was collision-induced dissociation C-trap (CID) with normalized collision energy (NCE) of 25 V, acquired over an m/z range of 350–2000. In the AIF scan, the mass resolution was 70,000 FWHM at 100 m/z with AGC target at 2×10^5 maximum IT 100 ms. Instrument control, data acquisition, and evaluation were done with the Q-Exactive Tune 2.1, Aria 1.3.6, and Thermo Xcalibur 2.2 software, respectively.

2.12.4. Analysis of Proteomic Data

Raw MS/MS data were analyzed using Proteome Discoverer 2.4.0.305 (Thermo Fisher Scientific). MS/MS spectra were searched against a human FASTA-formatted database (SwissProt, v2017–10-25 with taxonomy Homo sapiens, 42,252 sequences) using the SEQUEST HT algorithm. All database search was performed using a precursor mass tolerance of ±20 ppm and a fragment ion mass tolerance of ±0.6 Da. Enzyme specificity was selected to trypsin. Minimal peptide length was set to six amino acids with a maximum missed cleavages value of 2. Database searches were performed with carbamidomethylation on cysteine as static modification and oxidation on methionine and acetylation of protein N-terminal as possible modifications. For peptide and protein identification, the false discovery rate (FDR) with a target-decoy strategy was set to 0.01.

2.12.5. Statistical Analysis

Data presented are the mean ± standard error of the mean (SEM). Statistical evaluation was performed using STATISTICA data analysis software (ver. 13; StatSoft Inc., Tulsa, OK, USA). The Student's paired t-test was used to compare the differences between treated and untreated cells. A p-value of <0.05 was considered significant.

3. Results

3.1. OGT Expression and O-GlcNAc Level Are Glucose-Dependent in Mitochondria of Breast Cancer Cells

Previous observations indicate that global cellular *O*-GlcNAc level and expression of enzymes associated with this process are strongly regulated by glucose availability [15,26,27]. Since the recent data concerning the role of ncOGT and mOGT in mitochondrial protein *O*-GlcNAcylation are ambiguous, we first verified the expression of both splice variants in breast cancer cells cultured in different glucose conditions. To separately detect sequences encoding mOGT or ncOGT, we designed specific sets of primers. Using a RT-qPCR method, along with standardization against two housekeeping genes, we observed that the level of mRNA encoding both mOGT and ncOGT in breast cancer cells were significantly decreased in high glucose concentration (Figure 1B). The strongest impact of glucose concentration on *OGT* gene expression level was monitored for Hs578t cells. Each cell lines expressed mOGT mRNA at levels approximately 5–10 fold lower compared to those encoding ncOGT. However, mRNA levels encoding both isoforms seem to be affected in the same way by glucose content.

Mitochondrial localization of mOGT was previously detected by Love et al. [28] in HeLa cells by using Western blot, although no data supports this observation in other human cell lines [19]. Since glucose impacts on OGT mRNA levels, we assessed mOGT and ncOGT expression at the protein level in mitochondria-enriched fraction and cytoplasmic fraction of cells grown for 72 h in low (2 mM) and high (25 mM) glucose concentrations (Figure 1C). As reported earlier [16,28], we detected a ~103 kDa immunoreactive band in the mitochondria fraction, whereas a ~117 kDa band corresponding to ncOGT was only observed in the cytosolic fraction (Figure 1C). The expression of the ~103 kDa isoform was inversely correlated to glucose concentration; on the other hand, the cytosolic ncOGT expression seems to be only slightly down-regulated by high glucose in MCF-7 and MDA-MB-231 cells, with no difference observed in Hs578t cells. Interestingly, in opposition to the glucose-dependent decrease in mOGT expression, the level of *O*-GlcNAcylated mitochondrial proteins was higher in the high glucose condition when compared to the low glucose concentration. We then monitored the OGA expression in the two fractions, the localization of this enzyme in mitochondria having been recently confirmed [18]. Even though we detected OGA both in mitochondrial and cytoplasmic fractions, the level of this enzyme was unchanged under the different glucose conditions. Together, these observations suggest that glucose availability regulates OGT level in mitochondria, likely by control of gene expression. The increased *O*-GlcNAc level in mitochondria despite the decreased level of OGT may be due to changes in OGA and OGT catalytic activity or increased UDP-GlcNAc level.

3.2. mOGT Up-Regulation in Breast Cancer Cells Increases Mitochondrial O-GlcNAc Level

To better understand the role of mOGT in the regulation of mitochondria activity and bioenergetics and to identify mOGT protein partners in breast cancer cells, we designed plasmids that encode full-length mOGT and a mOGT mutant lacking the second catalytic domain (CDII). These constructs were cloned into commercial vectors that enable their expression in breast cancer cells as fusion proteins with HaloTag located at the C-terminus (mOGT-HaloTag and ΔCD-mOGT-HaloTag; for the full-length and the catalytic dead isoforms, respectively). Initially, we tried to generate stable cell lines by exerting a selective pressure with G418. However, similarly to another group [29], we could not generate stable cell lines overexpressing active mOGT, this was probably due to mOGT cytotoxic effect. Therefore, our study was conducted using transient expression of mOGT-HaloTag or ΔCD-mOGT-HaloTag proteins *versus* only HaloTag protein as a control (Figure 2A). As shown in the Figure 2B, mOGT-HaloTag, ΔCD-mOGT-HaloTag proteins, and HaloTag alone were targeted to the mitochondria-enriched fraction. As expected, in cells transfected with pmOGT-HaloTag, the level of *O*-GlcNAcylated proteins in mitochondria was increased in contrast with cells transfected with *p*-ΔCD-mOGT-HaloTag, confirming the catalytic activity of mOGT-HaloTag in mitochondria. The results of densitometric analysis are shown in Figure S1. While levels of endogenous mOGT were dependent upon glucose concentrations (Figure 1C), the expression of mOGT-HaloTag in mitochondria-enriched fractions is independent of glucose availability (Figure 2C). Therefore, these findings suggest that mOGT expression in mitochondria is more likely dependent upon gene expression rather than proteostasis.

3.3. *Increased mOGT Expression Affects Mitochondrial Homeostasis and Cellular Energy Metabolism*

According to published data, the increase in *O*-GlcNAcylation of mitochondrial proteins contributes to impaired mitochondrial function. More precisely, *O*-GlcNAcylation regulates oxidative phosphorylation at multiple sites of the respiratory chain contributing to alterations of the mitochondrial membrane potential and ATP production rates [15–17]. To find out whether mOGT expression and its catalytic activity influence the mitochondrial metabolic state of breast cancer cells, the OGT expression in cells grown in normal glucose conditions was up-regulated by transfection with plasmids encoding either mOGT or catalytic dead version versus the vector only encoding HaloTag as a control. Firstly, we determined the impact of increased-mOGT expression on the mitochondrial membrane potential ($\Delta\Psi$m), which is considered an index of mitochondrial condition. The transmembrane potential was assessed by staining cells with JC-1, a potentiometric fluorescent dye that accumulates in energized mitochondria. To confirm the sensitivity of the JC-1 dye to the detection of changes in membrane potential, the CCCP control was used. Cells transfected with pmOGT-HaloTag showed a significant increase in mean JC-1 fluorescent shift ratio (in a range from 24% for MCF-7 to 8% for Hs578t cells) compared with the control (pHaloTag transfected) cells (Figure 3A). The mitochondrial membrane potential is the main bioenergetic parameter that controls respiratory rate. Respiration of mitochondria is regulated by both the membrane potential as well as by ATP, which directly inhibits cytochrome C oxidase in a $\Delta\Psi$m-independent way. To determine whether changes in the respiration rate were associated with mitochondrial ROS production, we stained the cells transfected with DNA vectors with the red fluorescent cell-permeable superoxide indicator MitoSOX. We observed that in all cell lines, the increase in the full-length or catalytically inactive mOGT expression led to a significant increase in mitochondrial superoxide levels (Figure 3B). In cells up-regulating mOGT, the marked increase in ROS levels was accompanied by an increase in mitochondrial membrane potential. The increase in mitochondrial membrane potential and the generation of intramitochondrial ROS are the common features of the early stage of apoptosis [30].

In order to determine the influence of mOGT on the energy state of breast cancer cells, we assayed the intracellular levels of ATP in cells transfected with pmOGT-HaloTag or pCD-mOGT-HaloTag *versus* pHaloTag and grown under different glucose concentrations. For the three breast cancer cell lines, expression of either mOGT-HaloTag or pCDmOGT-HaloTag caused a decrease in intracellular ATP levels in comparison to pHaloTag treated cells (Figure 3C). The most significant differences in ATP levels were observed for MDA-MB-231 cells. A decrease in ATP levels in cells was observed independently of the concentration of glucose.

Together, these observations suggest that the up-regulation of mOGT limits the mitochondrial ATP production and increases the generation of reactive oxygen species, which are probably responsible for the toxic effects of mOGT.

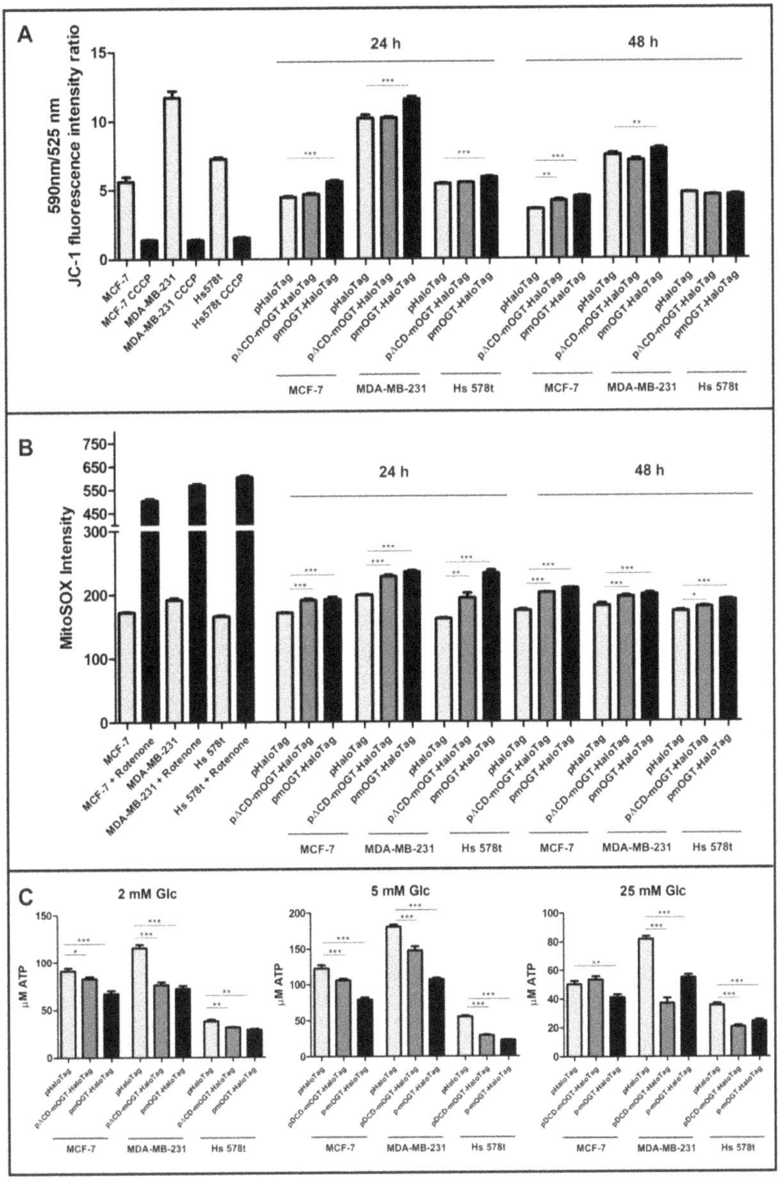

Figure 3. Elevated mOGT level impairs mitochondria homeostasis and ATP production. (**A**) Determination of mitochondrial transmembrane potential 24 or 48 h after transient transfection of cells with plasmid DNA. The 590/525 nm fluorescence intensity ratio of JC-1 dye is presented as a mean +/− SD. (**B**) Mitochondrial ROS level in transfected cells assessed by using MitoSOX Red mitochondrial superoxide indicator. The bars are shown as a median +/− SD of fluorescence intensity of cells determined by flow cytometry. (**C**) Intracellular ATP contents in cells treated for 48 h with pHaloTag (control), pΔCDmOGT-HaloTag, or pmOGT-HaloTag were determined using a Luminescent ATP Detection Assay Kit. (Data represent the average of at least 3 independent experiments performed in tetraplicate. * indicates significance $p < 0.05$; ** $p < 0.01$; *** $p < 0.001$. CCCP is a chemical agent which allows for complete dissipation of the proton gradient across the mitochondrial inner membrane.

3.4. mOGT Interacts and O-GlcNAcylates Multiple Sets of Mitochondrial Proteins

According to our results, mOGT has its own catalytic activity and is able to O-GlcNAcylate a myriad of mitochondrial proteins. To identify proteins interacting with the mOGT, we employed a HaloLink® Resin to co-purify the mOGT-HaloTag fusion protein with its mitochondrial interacting partners (HaloTag alone was used as negative control; see Methods). Transient transfections and pull-down efficiency were verified by Western blotting (Figure 4A,B). The enriched proteins were identified by nano-LC–MS/MS as described in the Methods. The candidate proteins interacting with mOGT were selected by a comparative analysis of the mOGT-HaloTag versus HaloTag protein partners. We identified many candidates interacting specifically with mOGT in breast cancer cells (Tables S1 and S6). Some of them were identified only in one cancer cell type, but 46 proteins were found in at least two of the three analyzed cell lines (Figure 5A,a). In the second approach, we intended to find out which mitochondrial proteins are O-GlcNAcylated by mOGT. Since O-GlcNAcylation of mitochondrial proteins can involve both mOGT and ncOGT, we used the RNA interference method to specifically knock-down ncOGT and increase the expression of mOGT by transient transfection. The sincOGT duplexes matched to the ncOGT mRNA sequence encoded by exon 2 of *OGT* gene; therefore, they specifically silenced only the long OGT isoform. In addition, the sequence of the mOGT-expressing vector was resistant to RNA interference. The efficiency of siRNA duplexes was checked by staining cells with anti-OGT antibodies and by visualization of the cells by confocal microscopy (Figure 4C). The efficiency of siRNA and pmOGT-HaloTag co-transfection on OGT and O-GlcNAcylation levels in the nuclear, mitochondrial, and cytoplasmic fractions of the three breast cancer cell lines was verified by Western blot (Figure 4A). Of particular interest, we noted that ncOGT silencing had no significant impact on OGT expression and O-GlcNAcylation levels in the mitochondrial fraction. On the other hand, as expected, a marked decrease in OGT expression and O-GlcNAcylation level was observed in the cytoplasmic and nuclear fractions. These observations suggest that ncOGT does not play a major role in O-GlcNAcylation of mitochondrial proteins in breast cancer cells.

Isolation of O-GlcNAcylated proteins from the mitochondrial-enriched fractions of cells previously transfected with sincOGT/pΔCD-mOGT-HaloTag, sincOGT/pmOGT-HaloTag or sincOGT/pHaloTag as a negative control was performed using the well characterized anti-O-GlcNAc antibody RL2 and the Pierce® Crosslink Immunoprecipitation Kit. Isolated proteins from O-GlcNAc-enriched probes were identified by nano-LC–MS/MS (see Methods). A previous study regarding the structure of human OGT in complex with a peptide substrate reported that the UDP moiety of UDP-GlcNAc binds in a pocket in the C-Cat domain (CDII) near the interface with the N-cat domain (CDI) containing two helices, which form an essential part of the enzyme active site [31]. In our study, the cells transfected with pΔCD-mOGT-HaloTag had slightly lower levels of O-GlcNAc modified proteins in mitochondria than the cells expressing HaloTag. We assume that this observation is associated with competition for binding to protein substrates between catalytic inactive ΔCD-mOGT-HaloTag and endogenous mOGT. Therefore, the mOGT-mediated O-GlcNAcylation of protein candidates was selected by a comparative analysis between samples that were co-transfected with sincOGT/pmOGT-HaloTag to those co-transfected with sincOGT/pΔCD-mOGT-HaloTag.

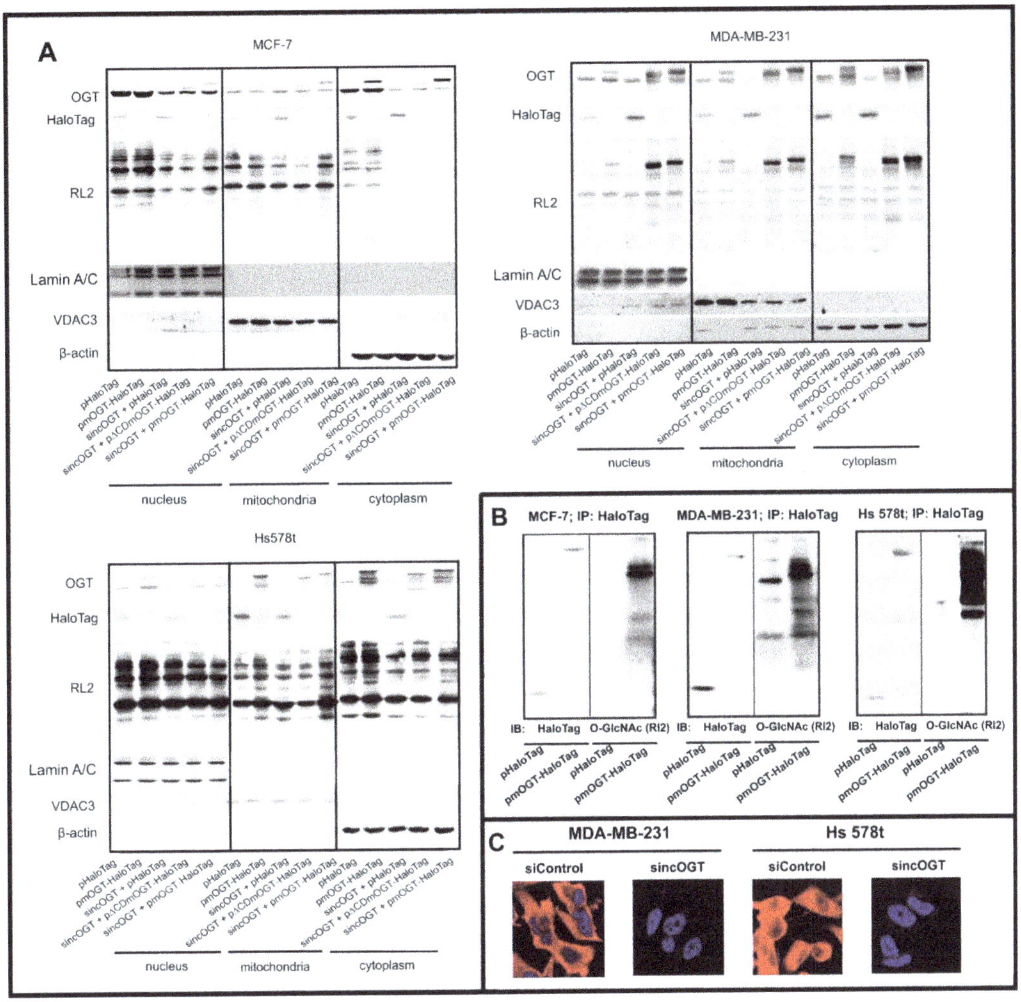

Figure 4. Isolation of mOGT candidate interacting proteins and modified substrates by proteomics approaches. (**A**) Representative Western blots present the effect of ncOGT knock-down and mOGT over-expression on the OGT and *O*-GlcNAc modified protein levels in nuclei- and mitochondria-enriched fractions and cytoplasm extracts from breast cancer cells. Blots were re-probed with Lamin A/C, VDAC3, and β-actin antibodies to verify the purity of isolated fractions. The mitochondrial lysates present herein were used for isolation of mOGT interactors or proteins modified by mOGT. (**B**) The confocal images of paraformaldehyde-fixed cells immunolabelled for OGT (red) counterstained with DAPI (blue) showing efficiency of ncOGT RNA interference. (**C**) mOGT-interacting proteins isolated from mitochondria-enriched lysates by using HaloLink Resin were probed with HaloTag and *O*-GlcNAc antibodies. (The original western blot images are shown in the File S1).

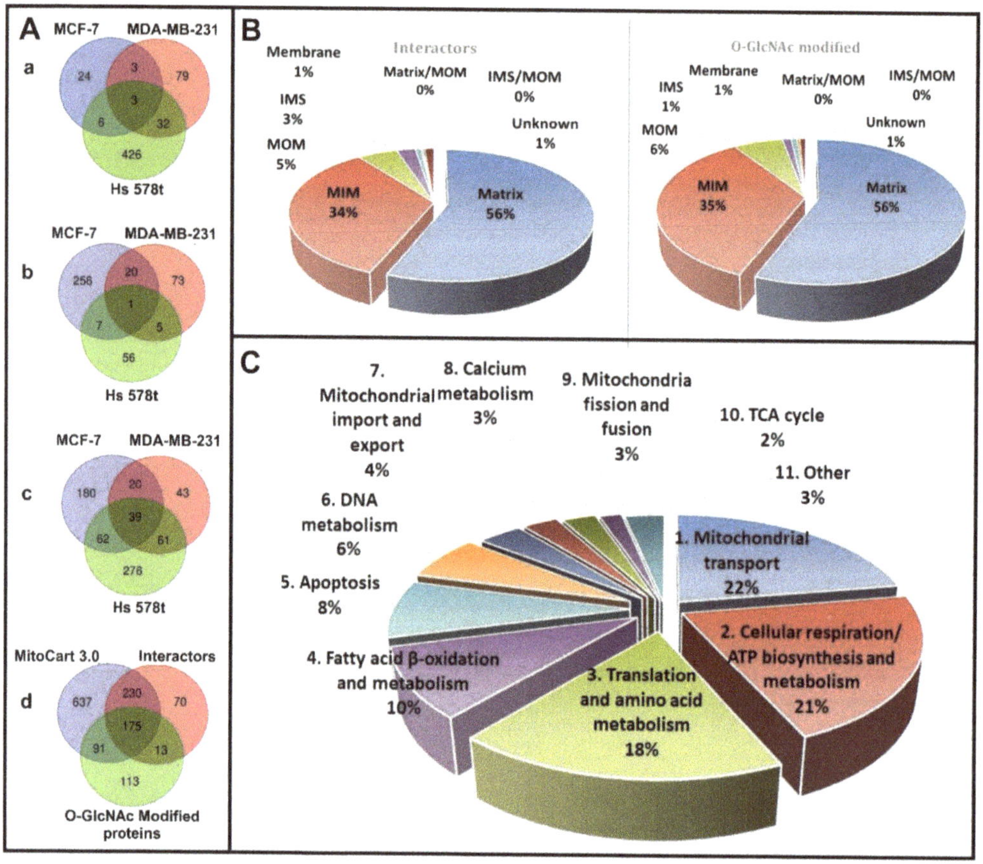

Figure 5. Proteomic analysis of isolated mOGT interactors and *O*-GlcNAc-modified candidates from mitochondria-enriched lysates of three analyzed cell lines. (**A**) Venn diagrams showing a, the total number of identified mOGT mitochondrial interactors; b, the total number of identified mOGT modified mitochondrial candidates; c, the total number of all identified mOGT protein partners and substrates; d, the interactors and *O*-GlcNAcylated candidates common with mitochondrial proteins according to MitoCart 3.0 datasets (Broad Institute). (**B**) The graph presents submitochondrial localization of identified interactors and *O*-GlcNAc modified candidates according to MitoCart 3.0 datasets. (**C**) The graph presents identified mitochondrial mOGT candidate proteins distributed into biological functional classes. MOM, mitochondrial outer membrane; IMS, intermembrane space; MIM, mitochondria inner membrane.

In the second step, all proteins identified in mitochondria of cells expressing only HaloTag were excluded. In this way, we identified a set of 33 *O*-GlcNAcylated candidates potentially modified by mOGT in at least two of the three tested cell lines (Figure 5A,b; Table S2). Moreover, some of these *O*-GlcNAcylated proteins were also identified as interacting partners of mOGT by using HaloLink Resin (Figure 5a,c; Table S3). We compared the identified in our study mOGT interactors and substrates with known mitochondrial proteins according to MitoCart 3.0 datasets (Broad Institute). We found out that 23% (266/1136) of mitochondrial proteins are potentially *O*-GlcNAcylated (Figure 5A,d) and over 35% of them may be interactors with mOGT. Most of these *O*-GlcNAcylated protein candidates are localized in the mitochondrial matrix and inner mitochondrial membrane (Figure 5B) and participate in mitochondrial transport, mitochondrial respiration, translation, fatty acid metabolism, apoptosis, and mtDNA processes (Figure 5C), making mOGT-managed mitochondrial *O*-GlcNAcylation a pivotal process affecting mitochondrial homeostasis. To narrow the list of mOGT interactors and substrates, we prepared a list of 107 proteins that fulfilled the following criteria: (1) were present in at least two of three cell lines, (2) were identified as mOGT interactors, and (3) were identified as O-GlcNAcylated proteins (Table S4). Of these 107 proteins, 53 were earlier identified as O-GlcNAc modified according to the Human O-GlcNAc database [32].

3.5. mOGT May Modify Some Nuclear Proteins

The study by Love and collaborators [28] showed that mOGT occurs mainly in the mitochondria thanks to the presence of a mitochondria-targeted sequence in the protein structure. Surprisingly, in our study, we found a set of nuclear proteins as potential mOGT partners or protein substrates. This was not due to contamination of mitochondrial fraction, because we did not detect lamin A/C in our mitochondria-enriched lysates.

The nuclear mOGT's putative substrates enriched by using both HaloLink Resin and anti-O-GlcNAc antibodies (RL2) are presented in Supplemental Table S5. The common candidates with nuclear localization were as follows: Isoform 4 of Apoptotic chromatin condensation inducer in the nucleus, Isoform AML-1L of Runt-related transcription factor 1, Isoform 10 of Transcriptional repressor CTCFL, NK1 transcription factor-related protein 2, and Isoform 2 of Chromodomain-helicase-DNA-binding protein 1.

4. Discussion

It has long been known that modification of proteins by O-GlcNAcylation controls target protein functions at many levels including subcellular trafficking, stabilization, complex formation, or enzymatic activity, and subsequently modulating metabolic and signaling networks [6]. O-GlcNAcylation level is tightly dependent upon HBP flux, which is supplied by diverse nutrients sources such as sugars, fatty acids, and amino acids. Thus, the O-GlcNAcylation level may be considered a cellular nutrient sensor [6,33]. Enhanced glucose flux through HBP results in increased O-GlcNAcylation. Therefore, since cancer cells metabolism requires accelerated glucose uptake and utilization, this results in an elevated rate of glycolysis and pentose phosphate pathway as well as O-GlcNAcylation processes [34]. Thereby, O-GlcNAcylation plays a pivotal role in cancer cell signaling and metabolism reprogramming. Champattanachai et al. [12], by using two-dimensional O-GlcNAc immunoblotting and LC–MS/MS compared O-GlcNAcylation of proteins in breast cancer tissue and normal tissue. They identified 29 proteins with O-GlcNAcylation differences, with seven being uniquely O-GlcNAcylated in breast cancer. Of these identified proteins, some were related to the Warburg effect, including metabolic enzymes, proteins involved in stress responses, and biosynthesis [12]. Until now, the research on links between O-GlcNAcylation and cancer biology has been more heavily concentrated on nuclear and cytoplasmic processes, whereas the reports related to the role of this modification in mitochondria are very limited. Moreover, the role of mOGT in O-GlcNAcylation of mitochondrial proteins remains insufficiently understood. To date, the mOGT expression has been reported only in the HeLa cell line [16,28]. In a recent study performed by Trapannone and collaborators [19] on the cell lines IL1R, HEK 293, Jurkat, SH-SY5Y, RAW, U2OS, and A549, surprisingly, mOGT protein was not detected. The authors concluded that this particular OGT isoform may be expressed only transiently, under specific conditions, and ncOGT is sufficient for O-GlcNAcylation of mitochondrial proteins. Therefore, as a starting point of this work, we verified mOGT expression in three breast cancer cell lines. For each cell line, the mRNA level encoding mOGT was 5–10 fold lower than ncOGT mRNA, and both of them were reduced upon increase in glucose availability. The glucose-dependent mRNA expression of mOGT and ncOGT was correlated with their protein levels in mitochondria and cytoplasm, respectively. Our results showed that, in breast cancer cells grown in different glucose conditions, mOGT level seems to correlate with expression of ncOGT. Thus, we suppose that mOGT is rather produced as a splice variant of OGT mRNA than driven by an independent promoter of the *OGT* gene. Our study is the first to report on the glucose-dependent expression of mOGT, probably through the effect of glucose on the regulation of *OGT* transcription. In turn, the level of O-GlcNAc-modified proteins in mitochondria was closely related to the availability of glucose and inversely correlated with mitochondrial OGT level. These observations are in agreement with previously published data, which showed that cardiac myocytes exposed to high glucose levels induced increased *O*-GlcNAcylation of mitochondrial proteins [15]. According to the literature, most mitochondrial proteins are synthesized in the cytoplasm, and they might be modified by ncOGT before targeting the mitochondria [19]. Thus, the loss of mOGT upon increased glucose levels might be compensated by ncOGT. Moreover, OGA activity can be lower, increasing the occupancy rate of *O*-GlcNAc on mitochondrial proteins. Nevertheless, in our study, the specific down-regulation of ncOGT had no significant impact on mitochondrial *O*-GlcNAc-modified protein level; thus, mOGT isoform may play a pivotal role in *O*-GlcNAcylation processes inside mitochondria and could intervene in reprogramming of mitochondria metabolism in breast cancer cells.

Actually, increasing evidence points to a critical role for *O*-GlcNAcylation in regulating mitochondrial function and cellular bioenergetics. To explore how increased mOGT expression and *O*-GlcNAcylation affects mitochondria and cellular energy metabolism, we designed and used a plasmid encoding the catalytically active fused mOGT protein and the catalytically inactive mutant displaying a deletion of the second catalytic domain. Our results showed that the increased mOGT expression caused a decrease in intracellular ATP level and altered glycolytic activity in comparison to control cells (Figure 3 and Figure S2). A similar effect associated with a decrease in glycolytic rate and oxygen consumption has been recently observed in SH-SY5Y and NT2 cells with elevated total and mitochondrial *O*-GlcNAcylation following treatment with TMG, a potent OGA inhibitor [35]. However, in a previous study, the authors noted that a decreased mitochondrial *O*-GlcNAcylation in SH-SY5Y cells by up-regulation of OGA expression also caused a significant decrease of oxygen consumption, glycolytic rate, and production of ATP [27]. Therefore, it seems that down-regulation and up-regulation of *O*-GlcNAc have a deleterious effect on mitochondrial function and energy metabolism. Interestingly, another study concerning the role of mitochondrial OGT isoform showed that a reduction of endogenous mOGT by siRNA in HeLa cells is associated with an increase in mitochondrial respiration; however, no significant differences in glycolytic rates were detected [16].

The first report regarding mOGT function indicated that mOGT overexpression in INS-1 cells triggers cell cytotoxicity and apoptosis [29]. Up-regulation of mOGT in breast cancer cells increases the mitochondrial membrane potential and the generation of intramitochondrial ROS, often observed in the early stage of apoptosis [30,36]. The same impact of mOGT on mitochondria membrane potential was also described in HeLa cells [16]. Our results argue that the toxic effect of mOGT is more likely related to inhibition of ATP synthesis in mitochondria and ROS production rather than calcium influx to mitochondria (Figure 3 and Figure S3). Similarly, in cardiac myocytes, increased mitochondrial *O*-GlcNAcylation induced by high glucose exposure was associated with impaired activity of complexes I, III, and IV of the respiratory chain in addition to lower mitochondrial calcium and cellular ATP content [15]. As expected, up-regulated mOGT in breast cancer cells had a toxic effect and led to the loss of cell viability independently of the concentration of glucose (Figure S4).

To shed more light on mOGT significance for mitochondrial activity, we took a proteomics-based approach to identify mOGT-interacting partners and mOGT-mediated *O*-GlcNAc-modified substrates. For the first time, we used the methodical approach that reduced the impact of ncOGT for the benefit of an increased probability of identifying the mOGT's protein partners and substrates. The results allowed us to specify a group of over 600 proteins as interactors and substrates for mOGT. The identified interacting proteins only partially overlap with the list of *O*-GlcNAcylated proteins from the same cell line. This may result from the way that the samples were prepared for each of the approaches, and especially that the information collected for each of them is different. Thus, protein–protein interactions are generally labile and transient, so many mOGT partners may not have been identified. A cross-linking strategy could help remedy this problem. The anti-*O*-GlcNAc antibody used, RL2, does not specifically recognize all *O*-GlcNAcylated proteins; thus many of the mitochondrial proteins modified by *O*-GlcNAc may not have been enriched and thus identified. Lastly, *O*-GlcNAcylation is also labile and unstable, since it can be quickly removed by OGA. On the other hand, a major positive point is that identified proteins were partially common between the different cell lines, reinforcing the specificity of our strategy. It should be noted that we did not expect a more significant coverage of candidates identified between the cell line because of the disparity of the proteomes. However, we indicated a group of 107 proteins that were present at least in two of three cell lines and were both identified as mOGT interactors and *O*-GlcNAc-bearing proteins. Many of these proteins have been previously proposed as *O*-GlcNAc-modified ones [16,27]. Our analysis revealed that mOGT interacts with and modifies the proteins participating in a variety of mitochondrial processes, such as transport, respiration, amino acid metabolism, protein translation, fatty acid metabolism, apoptosis, and mtDNA processes (Figures 5C and 6). Interestingly, despite the confirmed purity of the mitochondria-enriched fractions, we found a set of 73 candidates identified as nuclear proteins (Table S5). We cannot exclude the possibility that the presence of nuclear proteins among mOGT interactors is a result of mOGT leakage to the nucleus due to its over-expression. However, most of these proteins were identified by two different approaches: the HaloLink^{TM} Resin and the RL2 antibodies. Thus, they were both identified as interactors and *O*-GlcNAcylated proteins. Most of the characterized proteins are associated with chromatin condensation and DNA repair. A previous study by Lazarus and coworkers [20] reported certain nuclear-localized proteins as mOGT substrates. Interestingly, the characterized stretch of three amino-acid (DFP; residues 451–453) as the nuclear localization signal of ncOGT occurs in a region common to that of mOGT [37]. However, to date, the nuclear localization of mOGT has never been observed. Therefore, by analogy to mitochondrial proteins *O*-GlcNAcylated by ncOGT, we suppose that some nuclear proteins may be modified by mOGT before being directed to the nucleus, however, this hypothesis must be validated in the future.

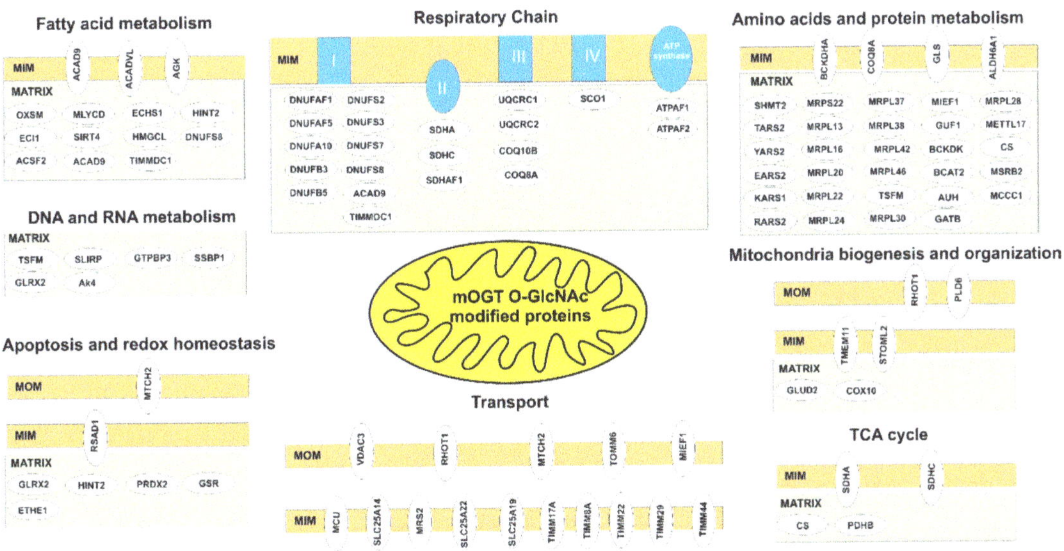

Figure 6. mOGT mediates *O*-GlcNAcylation of proteins involved in many processes central to mitochondria function. The circles indicate identified proteins characterized in Table S4 (Gold List). MON, mitochondria outer membrane; IMS, intermitochondria space, MIM, mitochondria inner membrane.

5. Conclusions

Protein *O*-GlcNAcylation is rapidly emerging as a key regulator of critical biological processes. In this study, we focused on the role of poorly explored mitochondrial OGT isoform. We found that the expression of mOGT in breast cancer cells is glucose-dependent. Increased mOGT expression affects mitochondria function and cellular bioenergetics in breast cancer cells. It seems that mOGT dysregulation may be responsible for changes in cellular energy metabolism via interaction with and *O*-GlcNAcylation of a large number of mitochondrial proteins that participate in many processes (Figure 6). In addition, our results suggest that mOGT modifies a specific set of nuclear proteins that deserves to be investigated in the near future.

Supplementary Materials: The following are available online at https://www.mdpi.com/article/10.3390/cancers13122956/s1. Table S1: mOGT Interactors, Table S2: *O*-GlcNAcylated candidates, Table S3: Both method per cell line, Table S4: Gold list, Table S5: Nuclear candidates, Table S6: All mitochondrial Candidates, File S1. The original western blot figures, File S2: Supplementary methods, Figure S1: Densytometric analysis of O-GlcNac levels in mitochondria-enriched fraction from Western blots presented on Figure 2B, Figure S2: mOGT up-regulation alter cellular respiration and glycolytic activity, Figure S3: Intracellular and mitochondrial calcium levels in breast cancer cells with up-regulated mOGT, Figure S4: mOGT treated cells lose the viability and change their cell cycle.

Author Contributions: Conceptualization, P.J.; methodology, P.J., P.C., P.K.Z., K.K., J.O., A.-S.V.-E., S.F., F.B., G.B. and D.Ż.; validation, P.J. and A.K.; formal analysis, P.J. and T.L.; investigation, P.J., P.C. and P.K.Z.; resources, J.O. and F.B.; data curation, P.J. and A.K.; writing—original draft preparation, P.J.; writing—review and editing, P.J., A.K. and T.L.; visualization, P.J. and P.C.; supervision, T.L. and A.K.; project administration, P.J.; funding acquisition, P.J. All authors have read and agreed to the published version of the manuscript.

Funding: This research was funded by the National Science Centre (NCN), grant number 2015/19/D/NZ3/01488. The research stay at University of Lille was founded by the Polish National Agency of Academic Exchange (NAWA), grant number PPN/BEK/2018/1/00328/DEC/1.

Institutional Review Board Statement: Not applicable.

Informed Consent Statement: Not applicable.

Data Availability Statement: The data presented in this study are available in this article (and Supplementary materials).

Conflicts of Interest: The authors declare no conflict of interest. The funders had no role in the design of the study; in the collection, analyses, or interpretation of data; in the writing of the manuscript, or in the decision to publish the results.

References

1. Pedersen, P.L. Warburg, me and Hexokinase 2: Multiple discoveries of key molecular events underlying one of cancers' most common phenotypes, the "Warburg Effect", i.e., elevated glycolysis in the presence of oxygen. *J. Bioenerg. Biomembr.* **2007**, *39*, 211–222. [CrossRef] [PubMed]
2. Hammad, N.; Rosas-Lemus, M.; Uribe-Carvajal, S.; Rigoulet, M.; Devin, A. The Crabtree and Warburg effects: Do metabolite-induced regulations participate in their induction. *Biochim. Biophys. Acta* **2016**, *1857*, 1139–1146. [CrossRef] [PubMed]
3. Hanover, J.A.; Krause, M.W.; Love, D.C. The hexosamine signaling pathway: O-GlcNAc cycling in feast or famine. *Biochim. Biophys. Acta* **2010**, *1800*, 80–95. [CrossRef] [PubMed]
4. Copeland, R.J.; Bullen, J.W.; Hart, G.W. Cross-talk between GlcNAcylation and phosphorylation: Roles in insulin resistance and glucose toxicity. *Am. J. Physiol. Endocrinol. Metab.* **2008**, *295*, 17–28. [CrossRef]
5. Chaiyawat, P.; Netsirisawan, P.; Svasti, J.; Champattanachai, V. Aberrant *O*-GlcNAcylated Proteins: New Perspectives in Breast and Colorectal Cancer. *Front. Endocrinol.* **2014**, *5*, 193. [CrossRef]
6. Kreppel, L.K.; Blomberg, M.A.; Hart, G.W. Dynamic glycosylation of nuclear and cytosolic proteins. Cloning and characterization of a unique O-GlcNAc transferase with multiple tetratricopeptide repeats. *J. Biol. Chem.* **1997**, *272*, 9308–9315. [CrossRef]
7. Gao, Y.; Wells, L.; Comer, F.I.; Parker, G.J.; Hart, G.W. Dynamic O-glycosylation of nuclear and cytosolic proteins: Cloning and characterization of a neutral, cytosolic beta-N-acetylglucosaminidase from human brain. *J. Biol. Chem.* **2001**, *276*, 9838–9845. [CrossRef]

8. Wani, W.Y.; Chatham, J.C.; Darley-Usmar, V.; McMahon, L.L.; Zhang, J. O-GlcNAcylation and neurodegeneration. *Brain Res. Bull.* **2017**, *133*, 80–87. [CrossRef] [PubMed]
9. Harwood, K.R.; Hanover, J.A. Nutrient-driven O-GlcNAc cycling-think globally but act locally. *J. Cell Sci.* **2014**, *127*, 1857–1867. [CrossRef]
10. Ferrer, C.M.; Sodi, V.L.; Reginato, M.J. O-GlcNAcylation in Cancer Biology: Linking Metabolism and Signaling. *J. Mol. Biol.* **2016**, *428*, 3282–3294. [CrossRef] [PubMed]
11. Krześlak, A.; Forma, E.; Baernaciak, M.; Romanowicz, H.; Bryś, M. Gene expression of O-GlcNAc cycling enzymes in human breast cancers. *Clin. Exp. Med.* **2012**, *12*, 61–65. [CrossRef] [PubMed]
12. Champattanachai, V.; Netsirisawan, P.; Chaiyawat, P.; Phueaouan, T.; Charoenwattanasatien, R.; Chokchaichamnankit, D.; Punyarit, P.; Srisomsap, C.; Svasti, J. Proteomic analysis and abrogated expression of *O*-GlcNAcylated proteins associated with primary breast cancer. *Proteomics* **2013**, *13*, 2088–2099. [CrossRef] [PubMed]
13. Jiang, K.; Gao, Y.; Hou, W.; Tian, F.; Ying, W.; Li, L.; Bai, B.; Hou, G.; Wang, P.G.; Zhang, L. Proteomic analysis of *O*-GlcNAcylated proteins in invasive ductal breast carcinomas with and without lymph node metastasis. *Amino Acids* **2016**, *48*, 365–374. [CrossRef]
14. Netsirisawan, P.; Chokchaichamnankit, D.; Saharat, K.; Srisomsap, C.; Svasti, J.; Champattanachai, V. Quantitative proteomic analysis of the association between decreasing O-GlcNAcylation and metastasis in MCF-7 breast cancer cells. *Int. J. Oncol.* **2020**, *56*, 1387–1404. [CrossRef]
15. Hu, Y.; Suarez, J.; Fricovsky, E.; Wang, H.; Scott, B.T.; Trauger, S.A.; Han, W.; Hu, Y.; Oveleye, M.O.; Dillmann, W.H. Increased enzymatic O-GlcNAcylation of mitochondrial proteins impairs mitochondrial function in cardiac myocytes exposed to high glucose. *J. Biol. Chem.* **2009**, *284*, 547–555. [CrossRef] [PubMed]
16. Sacoman, J.L.; Dagda, R.Y.; Burnham-Marusich, A.R.; Dagda, R.K.; Berninsone, P.M. Mitochondrial *O*-GlcNAc Transferase (mOGT) Regulates Mitochondrial Structure, Function, and Survival in HeLa Cell. *J. Biol. Chem.* **2017**, *292*, 4499–4518. [CrossRef] [PubMed]
17. Ma, J.; Liu, T.; Wei, A.C.; Banerjee, P.; O'Rourke, B.; Hart, G.W. O-GlcNAcomic Profiling Identifies Widespread O-GlcNAcylation in Oxidative Phosphorylation System Regulating Cardiac Mitochondrial Function. *J. Biol. Chem.* **2015**, *290*, 29141–29153. [CrossRef]
18. Benerjee, P.S.; Ma, J.; Hart, G.W. Diabetes-associated dysregulation of O-GlcNAcylation in rat cardiac mitochondria. *Proc. Natl. Acad. Sci. USA* **2015**, *112*, 6050–6055. [CrossRef]
19. Trapannone, R.; Mariappa, D.; Ferenbach, A.T.; Aalten, D.M.F. Nucleocytoplasmic human O-GlcNAc transferase is sufficient for O-GlcNAcylation of mitochondrial proteins. *Biochem. J.* **2016**, *473*, 1693–1702. [CrossRef]
20. Lazarus, B.D.; Love, D.C.; Hanover, J.A. Recombinant O-GlcNAc transferase isoforms: Identification of OGlcNAcase, yes tyrosine kinase, and tau as isoform-specific substrates. *Glycobiology* **2006**, *16*, 415–421. [CrossRef]
21. Bond, M.R.; Hanover, J.A. A little sugar goes a long way: The cell biologyof O-GlcNAc. *J. Cell Biol.* **2015**, *208*, 869–880. [CrossRef]
22. Reers, M.; Smiley, S.T.; Mottola-Hartshorn, C.; Chen, A.; Lin, M.; Chen, L.B. Mitochondrial membrane potential monitored by JC-1 dye. *Methods Enzymol.* **1995**, *260*, 406–417. [CrossRef]
23. Kockmann, T.; Trachsel, C.; Panse, C.; Wahlander, A.; Selevsek, N.; Grossmann, J.; Wolski, W.E.; Schlapbach, R. Targeted proteomics coming of age–SRM, PRM and DIA performance evaluated from a core facility perspective. *Proteomics* **2016**, *16*, 2183–2192. [CrossRef] [PubMed]
24. Geiger, T.; Madden, S.F.; Gallagher, W.M.; Cox, J.; Mann, M. Proteomic Portrait of Human Breast Cancer Progression Identifies Novel Prognostic Markers. *Cancer Res.* **2012**, *72*, 2428–2439. [CrossRef] [PubMed]
25. Velloso, F.J.; Campos, A.R.; Sogayar, M.C.; Correa, R.G. Proteome profiling of triple negative breast cancer cells overexpressing NOD1 and NOD2 receptors unveils molecular signatures of malignant cell proliferation. *BMC Genom.* **2019**, *20*, e152. [CrossRef] [PubMed]
26. Ngoh, G.A.; Facundo, H.T.; Hamid, T.; Dillmann, W.; Zachara, N.E.; Jones, S.P. Unique hexosaminidase reduces metabolic survival signaland sensitizes cardiac myocytes to hypoxia/reoxygenation injury. *Circ. Res.* **2009**, *104*, 41–49. [CrossRef]
27. Tan, E.P.; Villar, M.T.; Lezi, E.; Lu, J.; Selfridge, J.E.; Artigues, A.; Swerdlow, R.H.; Slawson, C. Altering *O*-Linked-*N*-Acetylglucosamine Cycling Disrupts Mitochondrial Function. *J. Biol. Chem.* **2014**, *289*, 14719–14730. [CrossRef] [PubMed]
28. Love, D.C.; Kochan, J.; Cathey, R.L.; Shin, S.H.; Hanover, J.A. Mitochondrial and nucleocytoplasmic targeting of O-linked GlcNAc transferase. *J. Cell Sci.* **2003**, *116*, 647–654. [CrossRef]
29. Shin, S.H.; Love, D.C.; Hanover, J.A. Elevated O-GlcNAc-dependent signaling through inducible mOGT expression selectively triggers apoptosis. *Amino Acids* **2011**, *40*, 885–893. [CrossRef] [PubMed]
30. Boren, J.; Brindle, K.M. Apoptosis-induced mitochondrial dysfunction causes cytoplasmic lipid droplet formation. *Cell Death Differ.* **2012**, *19*, 1561–1570. [CrossRef]
31. Lazarus, M.B.; Nam, Y.; Jiang, J.; Sliz, P.; Walker, S. Structure of human O-GlcNAc transferase and its complex with a peptide substrate. *Nature* **2011**, *469*, 564–567. [CrossRef]
32. Wulff-Fuentes, E.; Berendt, R.R.; Massman, L.; Danner, L.; Malard, F.; Vora, J.; Kahsay, R.; Stichelen, O.-V.S. The Human *O*-GlcNAcome Database and Meta-Analysis. *Sci. Data* **2021**, *8*, 25. [CrossRef]
33. Jóźwiak, P.; Forma, E.; Bryś, M.; Krześlak, A. O-GlcNAcylation and Metabolic Reprograming in Cancer. *Front. Endocrinol.* **2014**, *5*, 145. [CrossRef]
34. Hanover, J.A.; Chen, W.; Bond, M.R. O-GlcNAc in cancer: An Oncometabolism-fueled vicious cycle. *J. Bioenerg. Biomembr.* **2018**, *50*, 155–173. [CrossRef] [PubMed]

35. Tan, E.P.; McGreal, S.R.; Graw, S.; Tessman, R.; Koppel, S.J.; Dhakal, P.; Zhang, Z.; Machacek, M.; Zachara, N.E.; Koestler, D.C.; et al. Sustained O- GlcNAcylation reprograms mitochondrial function to regulate energy metabolism. *J. Biol. Chem.* **2017**, *292*, 14940–14962. [CrossRef] [PubMed]
36. Matsuyama, S.; Reed, J.C. Mitochondria-dependent apoptosis and cellular pH regulation. *Cell Death Differ.* **2000**, *7*, 1155–1165. [CrossRef] [PubMed]
37. Seo, H.G.; Kim, H.B.; Kang, M.J.; Ryum, J.H.; Yi, E.C.; Cho, J.W. Identification of the nuclear localisation signal of O-GlcNAc transferase and its nuclear import regulation. *Sci. Rep.* **2016**, *6*, 34614. [CrossRef]

MDPI
St. Alban-Anlage 66
4052 Basel
Switzerland
Tel. +41 61 683 77 34
Fax +41 61 302 89 18
www.mdpi.com

Cancers Editorial Office
E-mail: cancers@mdpi.com
www.mdpi.com/journal/cancers

www.ingramcontent.com/pod-product-compliance
Lightning Source LLC
Chambersburg PA
CBHW062104220926
51850CB00010B/2890
* 9 7 8 3 0 3 6 5 3 0 8 3 3 *